Movement and rhythms of the stars

Movement and Rhythms of the Stars

A guide to naked-eye observation of Sun, Moon and planets

Joachim Schultz

Floris
Books

Edited by Suso Vetter and John Meeks
Translated by John Meeks

First published in German under the title *Rhythmen der Sterne* by
Philosophisch-Anthroposophischer Verlag am Goetheanum in 1963
This translation first published in 1986
This edition published in 2008
Third printing 2018

British Library CIP Data available
ISBN 978-086315-669-4
Printed by Lightning Source

Contents

The planets

Translator's foreword

This book appeals so directly to the faculty of observation that a special introduction regarding the method of presentation might seem superfluous at first. Nonetheless, certain difficulties may be posed for English readers by references to the Goetheanistic, or 'phenomenological' method. These two terms are generally used in the same sense, and the attempt may be allowed briefly to characterize them.

As used by Plato, the Greek word *phainomenon* means 'that which appears to the senses', in contrast to the *noumenon*, the underlying spiritual reality. It is therefore sometimes also expressed with the German word *Erscheinung*, meaning 'appearance'. The word 'apparent', when used to describe the motions of the luminaries as we directly perceive them, must therefore be understood in this sense, and not in the sense 'illusory', as is so often the case in astronomical terminology. We shall not be concerned to reduce the 'apparent' motions to 'real' ones, but rather to enter as fully as possible into the particular form-language of the celestial *phenomena* as they unfold above and around us.

It was Johann Wolfgang von Goethe (1749–1832) who gave the most enduring impulse to a phenomenological method in natural science. In his contributions in the fields of botany, zoology and the theory of colours, Goethe directed attention to the evaluation of the qualitative aspect of the sense-world. He did not look for mechanisms to explain organic forms, or postulate 'light-waves' to account for the different colours, but regarded form and colour as 'sensory-moral' (*sinnlich-sittlich*) qualities which cannot adequately be understood without the activity of a corresponding faculty in man.

The importance of Goethe's impulse was fully recognized by such thinkers as Novalis and Hegel in his own time; it was rejected, however, by academic science, which chose to pursue the path of increasing abstraction. Since that time much ridicule has been directed at Goethe's scientific efforts, while the greatest tribute is paid to his poetic genius. Goethe, however, regarded his own scientific works as of even greater significance and lasting value than his poetic ones.

Rudolf Steiner (1861–1925), whose complete edition and commentary of Goethe's scientific writings has laid an indispensable groundwork for all later scholarship in this field, rediscovered the genius of Goethe as a scientist and gave valuable impulses for a renewal of the different branches of natural science in accordance with the Goetheanistic method.

It is one of the tasks of the Goetheanum in Dornach, Switzerland, to encourage this renewal. The present work, originally published under the auspices of the

Mathematical-Astronomical Section at the Goetheanum, represents a beginning in this direction for the science of astronomy.

Although Goethe often took the opportunity to make astronomical observations through the seven-foot refractor at the university in Jena, he never found a direct access to astronomy as a science, as he felt its possibilities to be limited through the dependence on the telescope. The question might therefore be raised: is it at all justified to speak of a Goetheanistic astronomy? It must be understood that if we do so it is in the conviction that Goethe's *method* can be applied to the phenomena of astronomy with results as fruitful and significant as those which Goethe himself obtained in the fields of botany, zoology, colour-theory and so on. Accordingly, very few of the phenomena dealt with in this book are of a telescopic nature. The reader is stimulated to direct observation under the open sky, with the totality of phenomena before him at all times. What emerges is totally different from the picture of the cosmos given in most astronomy textbooks. In their breathing alternation between forward and retrograde movement, in the harmonious evolution of their loop-forms, and in the rhythmic pulsation of their variations in brightness, the planets appear to stand in intimate relation to the world of living form which was so important to Goethe. An active contemplation of these celestial rhythms can lead to a deeper intuitive understanding of the principles of form and development as we encounter them on the Earth.

This is a challenging and by no means easy task. The reader who wishes to supplement his study with regular observation will do well to acquire a rotating star map, such as the two-sided map *Zodiak* developed by Joachim Schultz and published at Dornach. Although the map is published in German, an English instruction booklet is available with the English names of the constellations. Also highly to be recommended is the annual *Sternkalender*, from the same publisher. Only a minimal knowledge of German is required in order to follow the movements of the planets throughout the year.

John Meeks

Introduction

The present work represents the fruits of more than two decades of work in the form of lectures, courses and essays, under the auspices of the Mathematical-Astronomical Section at the Goetheanum. Such an undertaking is not without its risks.

The goal has been to give an introduction to the phenomena of astronomy and an insight into the astronomical-terrestrial relationships, in which man is taken fully into account.

In contrast to the numerous and often in their own way excellent works now available, we shall not be primarily concerned with the *image* of astronomical reality in mathematical form. We shall attempt rather to find methods which can lead directly to the *experience* of this reality.

It lies in the method of this book that mathematics remains almost completely untouched. This should not be regarded as cause to disparage this aspect of astronomical knowledge; on the contrary, any treatment which aspires to be more than an approximate description of the astronomical phenomena can only be exact with the help of mathematical discipline. Many of the descriptions which follow are the direct result of mathematical investigations.

Historically, the development of mathematics is most intimately connected with that of astronomy. We must recognize, on the one hand, the decisive role which mathematics has played since ancient times in the cultivation and furtherance of astronomy, and of its emergence as a science altogether. Without the groundwork of mathematical thinking, modern astronomy would be unthinkable. On the other hand, mathematics has been stimulated, advanced and expanded by astronomy. Whole fields of mathematical thought owe their development to astronomy. Begotten and nourished at the well-springs of the cosmos, mathematics has contributed as a faithful servant of astronomy to the quantitative description of the astronomical phenomena, as well as to their conceptual grasp.

But mathematics has never been able to offer *more* than a conceptual grasp of the cosmos. And not even that to full extent, for the movements of the celestial bodies are only accessible to a limited extent to definitive calculation. The latter proves, as we shall see in the course of our study, inadequate with regard to a number of essential features of the cosmic reality.

Here we find an expression of the important fact that the cosmos does not bear the character of a mechanism which is fully calculable and comprehensible to the last details with mathematics and geometry. It reveals itself rather as an organism with its own development and dynamically changing rhythms. It is involved in a process of evolution which only during the present phase of its development has descended largely into the sphere of the calculable.

Mathematics therefore has the greatest significance for the gradual attainment of an overall understanding of the separate phenomena of astronomy in regular, systematic form, as well as of the properties of space, of *quantitative* relationships. But it should not be forgotten that only one aspect, the external aspect of reality, comes to expression here. A study of qualities, of the inner nature and essence of things, is not possible with these methods.

Stimulated by the challenge presented in trying to understand the movements of the celestial bodies, mathematics has given a great boost to astronomical science; it has proved a suitable tool for the determination of quantitative relationships. A description of the *influences* involved must go beyond the mere mathematical conceptions to spiritual insights, as they have been disclosed by Rudolf Steiner in Anthroposophy.

Through the tendency to an exclusively mathematical treatment of astronomy, the colourful variety of the phenomena, with its direct appeal to our senses, fades to grey shadows. At the same time, the manifold world of the spirit, the *reality* of spiritual beings and their influences, is not taken into account. One remains indifferent in an intermediate realm and loses the reality of the astronomical events in their inner and outer manifestation.

It therefore appears as a necessity of our time to direct attention once more to the celestial phenomena in all their purity. The phenomenological treatment in the sense of a Goethean method attempts to isolate the archetypal phenomena. It arranges the phenomena according to their logical connections. In a purely phenomenological study the principle of metamorphosis plays a prominent role. The cosmos is an organism of light and movement, which unfolds before us in constantly changing rhythms. Impressive examples of the principle of metamorphosis are given in the laws of the eclipses and the planetary loops.

In following this method calmly and with leisure, one becomes aware of many previously disregarded facts. The book draws attention to many such relationships, some of which may seem self-evident, but which throw light on significant new aspects when they can be read as a part of the overall signature of the phenomena.

There is a broad tendency in modern times to depreciate the outward *appearance* to mere semblance. It is in the sense of our study to re-constitute the 'appearance' as a physiognomical gesture which has something to tell us. In the phenomenon is manifested the governing principle, in the sequence of phenomena are revealed character and activity. In other words, *that which is relative can be recognized and evaluated positively as a relationship*, for it is a reality which comes to expression in all connections between Earth and Man, cosmos and Earth, or cosmos and Man. It is our task to become aware of real cosmic influences as they are manifested in the sequences of phenomena.

In the mere formulae of an overall mathematical treatment, we are given the

INTRODUCTION

'general solution' of the planetary movements, from which the orbits of the individual planets can be derived as needed.

For our present considerations the main emphasis is placed on individual cases. For it is precisely the variations, the different forms of metamorphosis — *how* Saturn, Jupiter, Mars and so on, carry out the general formula of planetary movement — that reveal the individual features of their character.

This method of presentation has the natural result that the astronomical phenomena begin to speak, to become gestures, in which wisdom-filled order becomes visible. For this reason the attempt has been made to allow the *qualities* of the phenomena to emerge in the drawings as well.

The sequences of phenomena may be understood in a deeper sense when they are complemented in the light of the insights of Anthroposophy. In this way the cosmos appears as an organism, as an ordered whole which owes its existence to the activity of spiritual beings. Anthroposophy offers new insights into cosmology, of which a few aspects may here be outlined.

13

1. The phenomenology of astronomy

The visible phenomena of astronomy surround us on all sides: geometrically speaking, we are completely enclosed by them. They appear to us, moreover, in rhythmic sequence (daily, annually and various other periods). They therefore display aspects in space as well as in time.

We shall not be concerned here with a rigid, photographic rendering of the details, but rather with the pure phenomena, and their mutual interrelationships. These cannot be arranged and classified in an arbitrary fashion; they must be brought into a sequence which is objectively justified. The momentary appearance of the planets in their relative positions, their cycles of increasing and decreasing brilliance and their periods of visibility must all be thought of as part of an organism in time, which must be extended imaginatively into past and future stages if it is to be brought into a complete context. This is only possible through the faculty of human thinking.

With the need for placing the single phenomenon within its context as part of a greater whole is connected the fact that astronomy, as a science of the senses, can only be grasped through practice.

We experience the phenomena of astronomy in three stages:

light, optical impression, radiance, twinkling of the luminaries,

movement of the luminaries,

rhythm, as it is revealed through light and movement.

Under the two aspects of light and movement all the visible phenomena of the heavens may be pursued. In both, however, rhythms are manifested. The most striking of these underlie the changing spatial positions of the stars and planets relative to each other and to the Earth. Alongside these 'periods of movement', however, must also be mentioned the 'light periods': rhythms which are expressed in the changing brightness of the Moon, the planets and other variable luminaries.

Our mode of observing the heavens has a twofold character. Our immediate optical impression is that of a plane studded with innumerable points of light, which appear at first to be at rest. Only with the passage of time do we perceive movement and direction.

The Greeks directed their attention to the active side, to the movements. For two thousand years astronomy was devoted primarily to this dynamic aspect of the phenomena. At the beginning of the modern age this teaching became permeated by the concepts of terrestrial physics, and thereby passed over into *celestial mechanics*. The optical side of the celestial phenomena, the *light* itself, did not become an object of serious scientific study until the nineteenth century.

This new branch of science, *astrophysics*, is devoted, as it were, passively, receptively, to the light.

A science of the celestial rhythms as a higher, independent branch of knowledge is, however, completely lacking. Yet this is the very key to an effective understanding of cosmology and cosmogony. The rhythms, appearing through the interplay of time and space, are the ordering, formative principle of all spatial and temporal cycles.

Movements and rhythms of the fixed stars

2. The stars and their daily movements

Morning, midday and evening are distinguished by the rising of the Sun in the east, its culmination in the south and its setting in the west. If we follow the Sun in the course of its daily movement, we find that it describes an arc above and around us in the sky (Figure 1).

After the evening twilight has faded, the stars appear one after another in the night sky, in the sequence of their brightness. In the south, in the same portion of the sky which the Sun traversed during the course of the day, we become aware of a gradual movement of the stars, from left to right, from east to west. Each star thus describes an arc similar to that of the Sun (Figure 2). Because of the great multitude of stars in the sky, we can simultaneously observe some stars rising in the east, others setting in the west, and still others culminating in the south.

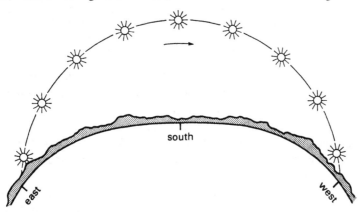

Figure 1. An arc described by the Sun over the horizon in the course of a day (spring or autumn). The arrow indicates the direction of movement from the east through the south towards the west.

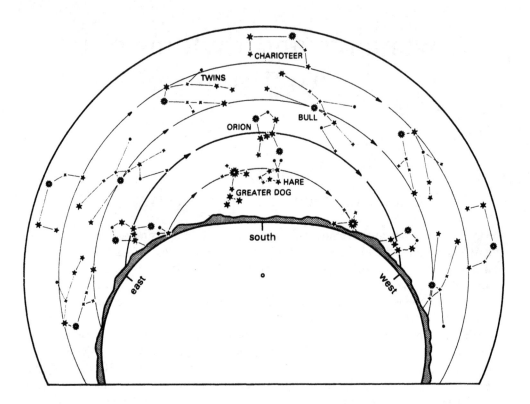

Figure 2. Movement of the stars over the southern horizon. The same constellations are shown in three positions. For instance, left: Dec 24 at 19.00; centre: Dec 24 at 24.00; right: Dec 25 at 5.00. Celestial equator: ————. The small circle beneath the south point of the horizon indicates the direction of the celestial south pole.

New groups of stars appear continually in the eastern sky, and climb obliquely to the right (southwards), while the constellations visible in the western sky sink obliquely towards the horizon and vanish.

The arcs of movement of the individual stars run parallel to one another. They may be distinguished through the fact that they rise to different heights in the south, thereby enclosing larger or smaller portions of the firmament (Figure 2).

In the north, too, we discover a left-to-right movement in the lower half of the sky, which in this case indicates a progression from west to east. Moreover, the arcs thus described continue upwards to form concentric circles around the celestial north pole. In the immediate vicinity of the latter stands the *Pole Star*, also called Polaris (Figure 3). In the northern hemisphere, the celestial pole is located directly above the north point of the horizon. Its angular height corre-

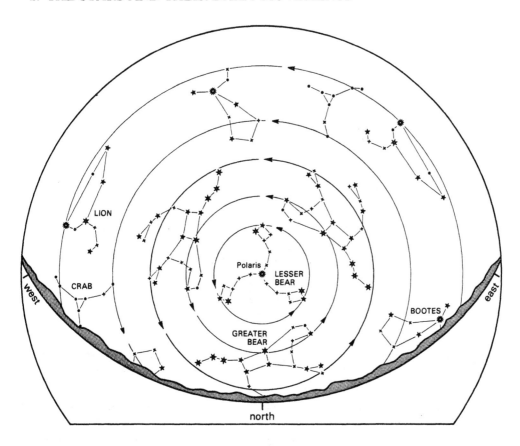

Figure 3. Movement of the stars over the northern horizon. The same constellations are shown in three positions. For instance, left: Sep 20 at 16.00; centre: Sep 20 at 24.00; right Sep 21 at 8.00. Boundary of circumpolar stars: ————.

sponds to the latitude of the observer. In London therefore, the Pole Star stands 51½° above the northern horizon, while in Aberdeen it is 57°, in America 30° (New Orleans), 40° (New York), 50° (Vancouver). The Pole Star, being very nearly stationary with respect to the horizon, enables us to determine the positions of all the cardinal points, and thus to orientate ourselves at a glance.

If we imagine the point diametrically opposite the celestial north pole, which, for the northern hemisphere, is always below the horizon, our gaze is led imaginatively through the Earth to the celestial south pole (Figure 4). It marks the centre of all the concentric circles of movement of the southern stars. It is to be imagined vertically below the south point of the horizon. Its angle below the horizon corresponds once more to the observer's geographical latitude. In Britain 50° to 57°, and in North America 30° to 50° below the horizon. The line connecting the

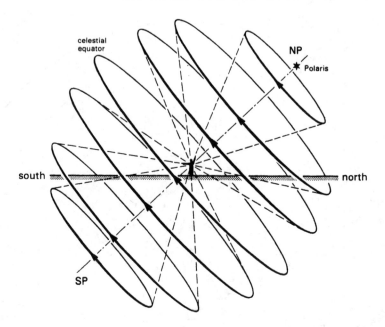

Figure 4. *Various cones which arise when the movements of the stars are followed with outstretched arms. The horizontal line represents the plane of the horizon. Celestial north pole: NP. Celestial south pole: SP. NP-SP is the celestial axis which runs through the observer.*

two celestial poles is the *celestial axis*; around it the whole star-sphere rotates once each day from east to west.

Figure 4 shows the relation of the daily circles of movement of the individual stars to the celestial axis. Let us suppose that an observer, facing either to the north or to the south, is able to visualize the position of the celestial axis. With outstretched arms he follows the various possible diurnal movements of the stars, thus 'drawing' them in the space around him. In following the paths of stars close to the pole, his arm describes a narrow cone. As the cones are allowed to become broader, they describe circles in the sky which correspond to the movements of stars further from the poles.

Because of the obliquity of the arcs of movement, those of the southern stars rise only a little distance above the horizon. The rising and setting points of all the circles lie symmetrically to either side of the south point of the horizon. As the cone broadens, the arc of movement rises higher. Finally, the cone flattens out into a plane at right angles to the celestial axis, and divides the entire sphere of the heavens into a northern and a southern hemisphere. The great circle of this plane is the *celestial equator* (equator = equalizer), separated by 90° from each pole.

In the northern sky, too, our understanding of the different circles of movement

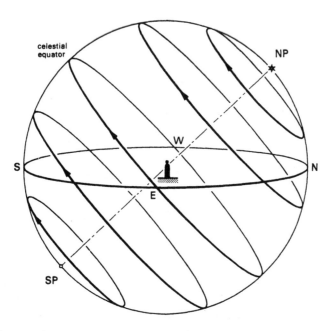

Figure 5. Spherical representation of the circles of daily movement of the fixed stars. N-E-S-W is the plane of the horizon.

is simplified by tracing their course with outstretched arms. Taking our starting point at the pole, we describe first narrower, then broader cones, until we reach the plane of the celestial equator. In this way we can visualize how each star progresses during the course of the night, and where we may find it at any given time.

Just as each point on the Earth has a distinct geographical latitude, ranging from 0° to ±90°, so too the positions of the individual stars on the celestial sphere are given by their angular divergence, or *declination* from the celestial equator, from 0° to ±90°. Thus, the Pole Star has an angular declination of almost 90° (+89° 07′); the stars in the belt of Orion in the immediate vicinity of the celestial equator have a declination of close to 0°, while Sirius has a (southern) declination of −17°.

In Figure 5 we see that of all the circles of movement, only the celestial equator is exactly half above and half below the plane of the horizon.* The stars and

---

*A distinction must be made between the *natural horizon*, which is formed by the local landscape with its mountains, valleys, trees, houses, and the *horizon-line*, which is formed by the intersection of the celestial sphere with the horizontal plane extending from the eyes of the observer. With respect to the celestial sphere, this coincides with the *mathematical horizon*, the plane of which passes through the centre of the Earth.

21

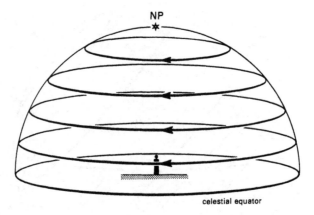

Figure 6. Circles of movement of the stars at the North Pole.

constellations near the celestial equator therefore move above the horizon for twelve hours, and then below it for twelve hours in the course of their daily movements. Since for all parts of the Earth the celestial equator always intersects the horizon exactly at the east and west points, it follows that all stars which rise exactly in the east and set exactly in the west must lie on the celestial equator.

Stars with a southern declination always remain below the horizon for more than twelve hours. With increasing northern declination, on the other hand, the circles of movement rise higher above the horizon. Those stars in the northern sky which never set are called *circumpolar stars*. Their circles of movement lie completely above the horizon; their circling motion around the Pole Star is especially clear (see Figures 3, 4 and 5).

Polar and equatorial regions

If the observer travels to the north or south, he will notice characteristic differences in the daily movements of the stars for the various geographical regions. We have already mentioned that the height of the Pole Star reflects the latitude of the observer. Thus the celestial north pole rises higher and higher as we travel northwards; at the same time the circles of diurnal movement intersect the horizon at smaller angles. The opposite takes place when we travel southwards: the celestial north pole sinks towards the northern horizon, and the paths of the stars intersect the horizon at steeper angles.

The *Arctic and Antarctic regions* are characterized by the very flat angles of rising and setting. Long, colourful twilight periods form the transition between day and night. At the Earth's North Pole the stars do not rise and set; they simply circle above the horizon, always retaining the same height (Figure 6). The celestial pole stands vertically above the observer, and the celestial equator coincides with

2. THE STARS AND THEIR DAILY MOVEMENTS

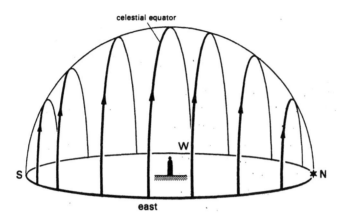

Figure 7. Arcs of movement of the stars at the earth's equator. N-E-S-W is the horizon.

the horizon. In the polar regions, therefore, the visible portion of the starry sky remains virtually the same throughout the year. At the North Pole only the northern celestial hemisphere can be seen, while at the South Pole only the southern hemisphere is visible.

For the observer on the *equator*, the celestial axis passes through the north and south points of the horizon. Here the vertical rising and setting of the stars is characteristic. This results in very brief twilight periods and rather abrupt transitions between day and night. The celestial equator passes from the east to the west point of the horizon, through the *zenith* of the celestial sphere, located vertically above the observer, and through the *nadir*,* located vertically beneath him (Figure 7).

Whereas a certain monotony prevails in the celestial movements at the poles, the tropics are characterized by the constantly changing face of the heavens. Within a period of twenty-four hours all the constellations of the sky rise above the horizon. Only the stars near the celestial poles, such as the Pole Star in the north and the faint constellation of the Octant in the south, remain for the most part shrouded by the haze close to the horizon.

The star phenomena in the *temperate zones* of the Earth occupy an intermediate position between the extremes of the poles and the tropics. Here there are stars which always circle above the horizon (the circumpolar stars); there are others which rise and set during the course of the day; and finally, there are stars which never appear above the horizon at all.

---

*The term *zenith* always denotes the direction vertically above the observer. The word is of Arabian derivation, and means 'direction of the head'. The point diametrically opposite the zenith in the celestial sphere, and therefore directly below the observer, is called the *nadir*, meaning 'opposite'.

The southern hemisphere

This is true for the southern hemisphere as well as the northern, with the fundamental difference that the apparent direction of the movements is reversed. A southern observer looking northwards sees the stars rising on his right hand in the east, culminating over the north point of the horizon and finally setting in the west, to his left. The circumpolar constellations in the south revolve in clockwise direction around the celestial pole.

It is a bewildering sight for an observant traveller who has just flown from one hemisphere to the other to find everything suddenly turning the 'wrong' way. There may be a feeling of disorientation, which is only the natural result of an unnaturally sudden change of location on the Earth's surface. In trying to understand how the reversal of direction comes about, a traveller overland or by boat is at a great advantage over an air-traveller, as he is able to observe the transitions from night to night.

Approaching the equator from the north he can see how the northern circumpolar stars become restricted to a smaller and smaller circle, shrinking finally into a point at the equator itself, where the celestial pole lies directly on the horizon. The stars which, in temperate latitudes, circled in anticlockwise direction around the Pole Star now describe semicircular arcs above the north point, rising in the east (right) and setting in the west (left). The movements of the stars in the southern sky are exactly symmetrical to those in the north, but in opposite direction: the diurnal movements proceed from left to right.

As the voyage continues southwards, the south celestial pole gradually rises above the horizon. The first bright constellation to become circumpolar is the Southern Cross, and closely following its circling movement the two brightest stars in the Centaur (the Pointers). The northern circumpolar constellations of the Little Bear and the Great Bear have by this time disappeared completely below the northern horizon. As in the northern hemisphere, the height of the celestial pole (ϕ) corresponds to the geographical latitude, and the height at which the equator culminates is the complementary angle ($90°-\phi$).

Figure 8 shows the movement of the stars as seen from a temperate southern latitude, looking north. Northern readers recognize the familiar constellations of Orion and the Bull, but they are the other way up. During the course of our imagined journey into the southern hemisphere these constellations will have risen higher and higher in the sky night after night. The Bull culminates directly overhead in the vicinity of the Tropic of Cancer, which lies at latitude $23\frac{1}{2}°$ N (Mexico, Egypt, India). This line demarcates the northernmost latitude at which the Sun can culminate in the zenith. Orion, which lies directly on the celestial equator, does not culminate overhead until the terrestrial equator is reached. Other constellations attain their zenith culminations at other geographical latitudes, depending on their northern or southern declination from the celestial equator. At the moment of a zenith culmination it is not possible to say that a constellation is

24

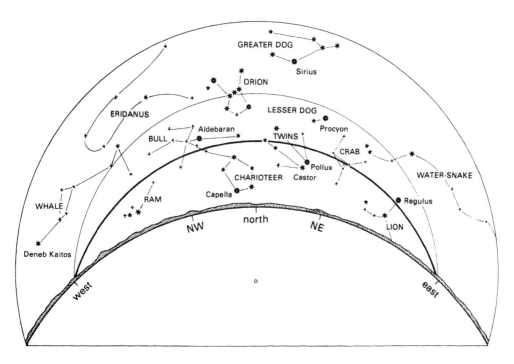

Figure 8. Stars of the southern hemisphere over the northern horizon, for instance on Dec 22 at 23.00.

orientated in a particular direction. Only before and after culmination does it appear to face first east, then west. What previously, in more northern latitudes, was an orientation of above and below, has now passed over into an orientation of right and left. As the journey proceeds southwards, the constellation culminates further and further to the north of the zenith, whereby a new orientation of above and below is attained, only in the reversed sense.

This touches upon one of the greatest obstacles experienced by inhabitants of the southern hemisphere in learning the constellations. All the major northern constellations bear names derived from the Greek myths; many are difficult to imagine pictorially, as they are the 'wrong' way up. The representations of these mythological figures on European star maps make no concessions to southern observers. The southern sky, on the other hand, is populated by a considerable number of constellations which are not the product of ancient mythological imagination, but of modern fantasy. They include the constellations of the Octant, the Telescope, the Air Pump, the Flying Fish, the Chameleon, the Peacock, the Toucan and the Indian Bird. These constellations, many of which are small and indistinct, contribute to the somewhat incongruous impression of the southern sky with its strange mixture of Greek mythological figures, exotic tropical fauna and

25

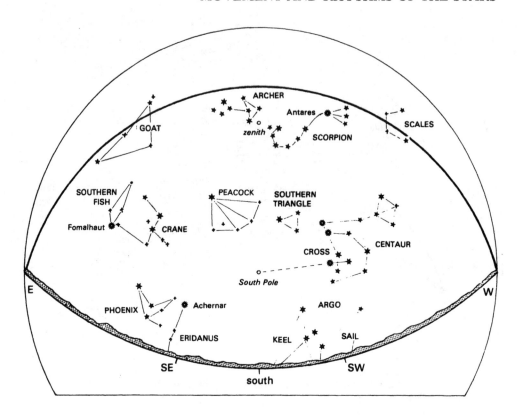

Figure 9. Stars of the southern hemisphere over the southern horizon, for instance on June 21 at midnight.

mundane scientific instruments. The relative difficulty which arises in obtaining a vivid pictorial grasp of the heavens can, however, serve as a challenge to seek in the individual constellations not so much realistic, static representations as pure forms permeated with dynamic qualities.

Figure 9 shows the view towards the southern horizon, with the brightest stars and constellations. As in Figure 3, the outermost circle represents the celestial equator. Both Figures 8 and 9 are drawn for the latitude 35° S (Sydney, Cape Town, Buenos Aires) as the inhabited areas in the southern hemisphere are generally closer to the equator than those in the northern hemisphere. (Figures 1 to 5 are drawn for 45° N.)

Sidereal time

The daily rotation of the celestial sphere, which is experienced differently in different parts of the Earth, is designated a *sidereal day*. At its completion the same stars once more reach their highest point, or *upper culmination* over the southern horizon.* For the astronomer the sidereal day provides the foundation for the exact measurement and determination of time. The length of the sidereal day can be calculated very precisely with the help of the telescope, from successive culminations, or *meridian passages*,† of the same star. The instrument used for these measurements, known as the meridian circle, is a telescope mounted on an east-west axis, so that it can only move up and down along the meridian. Measurements with such instruments show the *sidereal day* to be 23^h 56^m 04^s. The indication of time derived from the movement of the fixed stars is designated as *sidereal time*, or star time (ST). Its divergence from the full twenty-four hours will be discussed later.

Once we have determined the time and declination at which a particular star crosses the meridian, a basis has been established for defining its position in the celestial sphere. All stars which culminate at the same time must lie on the same celestial meridian. Those stars which, for example, culminate exactly 6, 12 or 18 sidereal hours later lie on celestial meridians which are separated from the first by 90°, 180° and 270°, respectively. All the celestial meridians intersect at the celestial poles. In this way the position of any star may be exactly defined, either through the indication of an angle, or through a corresponding measurement of time in hours, minutes and seconds. This value is called the *right ascension* of a star (AR, *ascensio recta*, 'straight ascent' from the celestial equator). The correspondence between time and angle is 24 hours = 360°, so 1^h = 15°.

As the starting point for the measurement of right ascension along the celestial equator, the spring equinox (First Point of Aries) has been chosen. This is the point where the Sun crosses the celestial equator each year around March 21. The meridian which passes through this point is designated as 0^h, or 0° (or 24^h, 360°).

The angle of a star vertically above or below the celestial equator is indicated by a specific *declination* (δ) either to the north (+) or to the south (−). This value is given in degrees, from 0° to ± 90°. Thus the position of a star is determined in relation to the celestial equator, and consequently to the whole star sphere, as soon as the numerical values of its right ascension and declination are known. These are given on most detailed star maps. A list can be found in the Appendix (Table 1.4).

Right ascension and sidereal time comprise a kind of cosmic clock. The division from 0^h to 24^h proceeds from west to east (anticlockwise); the 'clock-face' rotates from east to west (clockwise). When, on March 21 at 12.00 noon, the spring equinox culminates, the sidereal time is 0.00. Similarly, the meridian culminating at any other given moment determines the sidereal time. If the right ascension of

---

*Lower culmination is the lowest point of the arc over or under the north point of the horizon.

†The *meridian* is the great circle passing through the celestial poles and the north and south points of the horizon. All stars culminate on this line. The term may similarly be used to designate any circle which passes through both celestial poles.

a star is known (for instance Sirius, 6ʰ 44ᵐ), the sidereal time can be given directly at the moment of its culmination.

For an approximate calculation of local sidereal time, valid for the middle of each month at midnight, the following formula may be used:

$$2 \times m + 6 = ST \text{ (sidereal time)}$$

where m is the number of the month in question. Examples:

midnight in mid-April $2 \times 4 + 6$ $= 14^h$ ST

midnight in mid-November $2 \times 11 + 6 = 28^h = 4^h$ ST

For the beginning of the month subtract one hour; for the end of the month add an hour.

From the situation at midnight, the sidereal time can be calculated for any hour of the day. Assuming ST at midnight to be 3.00, it follows that ST 4 hours earlier at 20.00, was 23.00; at 5.00 it will be 8.00 ST.

The formula $2 \times m + 6 = ST$ is valid for midnight, 0° longitude, or for the longitude of a time zone (see Chapter 7). For 7½°E longitude it is $2 \times m + 5 = ST$. The formula can, of course, be adjusted for different times of day. For 21.00 it would be $2 \times m + 3 = ST$. On the basis of these principles, the formula can be corrected for every place and time of day.

In connection with the known right ascensions of a few stars, it should thus be possible at any given hour of the day or night to visualize the entire dome of the heavens with its star configurations. This is the principle upon which every rotating star-map is constructed.*

Apart from their popular names, the stars in each constellation were classified by Ptolemy with the Greek letters α, β, γ, and so on, in descending order of brightness. With changes in brightness, and accurate measurements, the order is no longer strictly accurate. Moreover, the stars are classified independently of their configurations into grades of brightness, or magnitude, indicated as mag 1, mag 2, and so on. The brightest and best known of the fixed stars belong to the classes mag −1, mag 0, and mag 1. Stars of mag 6 can just be seen on a clear night by an observer with good vision. The brightest fixed star, Sirius, has a magnitude of −1.5; the planet Venus can attain a luminosity of mag −4.5. The classes of magnitude do not follow a linear scale. The convention followed is such, that a star 100 times fainter than another has a magnitude 5 units greater. The individual steps therefore increase by a factor of $100^{-5} = 2.512$.

About 6 000 stars are visible to the unaided eye; with binoculars the number rises to about 60 000. A telescope having an objective lens of 10 cm brings the figure to over four million, and with the larger instruments ten million points of light come into view. Through the use of photography the number runs into the hundreds of millions.

*For a clearer understanding of these relationships the reader is referred to the *Drehbare Sternkarte Zodiak*, a rotating star-map designed by Joachim Schultz, published by Verlag Goetheanum, Switzerland.

3. The zodiac and its daily movement

As their name suggests, the fixed stars retain their relative positions in the sky. Sun, Moon and planets, on the other hand, continually change their positions against this constant background. The constellations through which these luminaries wander have always been regarded with special interest. They are the twelve constellations of the zodiac, and are designated as follows Ram (Aries ♈), Bull (Taurus ♉), Twins (Gemini ♊), Crab (Cancer ♋), Lion (Leo ♌), Virgin (Virgo ♍), Scales (Libra ♎), Scorpion (Scorpio ♏), Archer (Sagittarius ♐), Goat (Capricornus ♑), Water-Bearer (Aquarius ♒) and Fishes (Pisces ♓). They embrace the whole celestial sphere along the annual path of the Sun, dividing it into two hemispheres, as does the celestial equator. The Sun's path, called the *ecliptic*, is inclined at an angle of 23½° to the celestial equator, and intersects it at two diametrically opposite points in the sky. These are known as the *vernal* and *autumnal* equinoxes. (This nomenclature relates to the position of the Sun, which will be discussed further in the next chapter.)

Because of the inclination of the zodiac to the equator, its individual constellations describe quite different arcs during the course of a day. They culminate at different heights over the south point of the horizon, and rise and set at different points on the horizon (Figure 10).

Surrounding the points of intersection with the celestial equator lie the two constellations of the Fishes and the Virgin; their diurnal arc of movement falls together with the celestial equator. Both constellations, therefore, rise in the east and set in the west; they remain above the horizon for twelve hours, and below it for an equal length of time. This holds for all regions of the Earth, with the exception of the Poles.

Five constellations of the zodiac, namely the *Ram*, the *Bull*, the *Twins*, the *Crab* and the *Lion* stand above the celestial equator, in northern declination. Their diurnal arcs of movement rise to different heights in the southern sky. Their rising points lie between east and north-east, and their setting points between west and north-west. It is evident that this group of constellations remains above the horizon longer than below. This is most pronounced for the Bull and the Twins, which, in temperate latitudes, remain visible in the sky for about two-thirds of the 24-hour day, or roughly 16 hours. The opposite group of five constellations is situated to the south of the celestial equator: the *Scales*, the *Scorpion*, the *Archer*, the *Goat* and the *Water-Bearer*. They describe low arcs over the southern horizon with their rising points between east and south-east and their setting points between west and south-west. These constellations always remain longer below the horizon

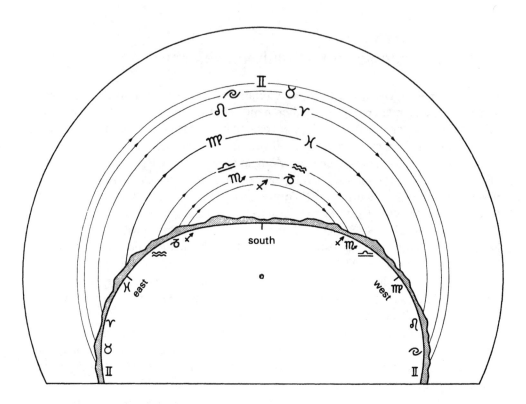

Figure 10. The diurnal arcs of movement of the twelve constellations of the zodiac. Celestial equator: ————; *the arrows indicate the direction of movement.*

than above. The southernmost constellations, the Scorpion and the Archer, appear for only about 8 hours in temperate latitudes.

The ancient division of the zodiac into the upper, or 'bright', and the lower, or 'dark' constellations reflects these relationships. The 'bright' constellations are those which for the greater part of the day send us their light directly; the 'dark' constellations, by contrast, are more often covered by the Earth. The difference is also apparent when we follow the diurnal arcs of the individual constellations

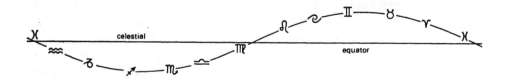

Figure 11. The position of the zodiac in relation to the celestial equator.

30

3. THE ZODIAC AND ITS DAILY MOVEMENT

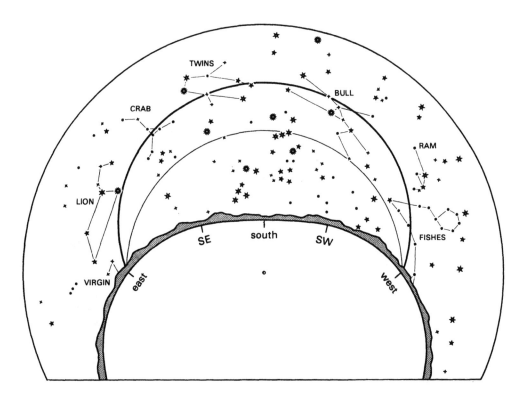

Figure 12. Steepest position of the zodiac, as on March 21 at 18.00, June 21 at 12.00, Sep 23 at 6.00, Dec 22 at 24.00. Celestial equator: ————. Sun's path (ecliptic) ————.

with our extended arm. For each of the upper constellations a cone is described which opens upwards; whereas those for the lower constellations open downwards (compare Figure 4).

The constellations of the Fishes and the Virgin occupy an intermediate position between these two groups. Here the cones flatten to a plane coinciding with the plane of the celestial equator. Also with respect to the duration of their visibility, these constellations assume a position of equilibrium.

The sequence and spatial arrangement of the zodiacal constellations is represented schematically in Figure 11. Their forms and relative sizes are apparent on the star map on Figure 25.

Figures 12 to 16 illustrate the changing positions of the zodiac over the horizon during the course of a day. At any given time exactly half the constellations of the zodiac arch across the southern sky. Figure 12 shows the zodiac as it appears just before the culmination of the Twins. The Bull, the Ram and the Fishes, having lower arcs of movement, are adjacent in the western part of the sky, while

31

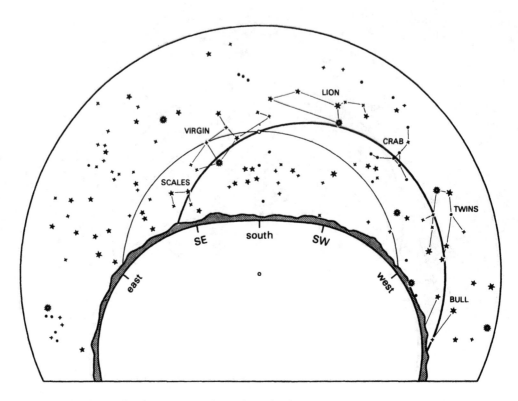

Figure 13. Intermediate situation of the zodiac, displaced to the west, as on March 21 at 24.00, June 21 at 18.00, Sep 23 at 12.00, Dec 22 at 6.00. Celestial equator: ———. Sun's path (ecliptic) ———.

in the east the Crab, the Lion and a part of the Virgin are rising along their respective arcs. Figures 13, 14 and 15 give the positions of the zodiac at further intervals of six hours. New constellations continue to rise in the east while others set in the west, so that different groupings appear over the horizon. Depending on whether the upper or lower constellations are visible in the sky, the zodiac rises alternately as a steep arc, high in the southern sky (Figure 12), or 12 hours later as a lower, flatter arc just over the southern horizon (Figure 14). In Figures 13 and 15 the intermediate positions are shown, in which the zodiac is displaced towards the west or the east.

The variation in the height of the zodiac at different times of day is determined by its inclination of 23½° to the celestial equator. The latter's height over the south point of the horizon stands in inverse relationship to the observer's latitude. Given a northern latitude of ϕ, the celestial equator will stand 90° — ϕ over the south point. For instance in London, latitude 51½° N., the height of the equator is 38½° (90° – 51½°). These spatial relationships are illustrated in Figure 5. In

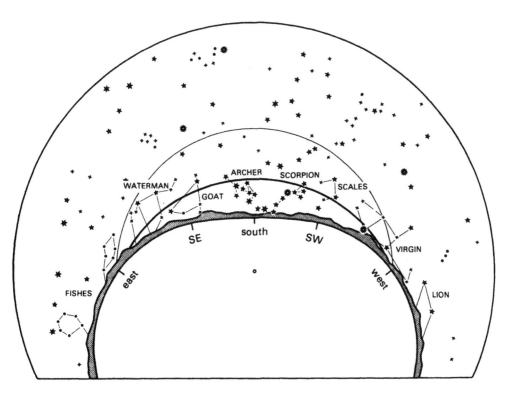

Figure 14. Lowest situation of the zodiac, as on March 21 at 6.00, June 21 at 24.00, Sep 23 at 18.00, Dec 22 at 12.00. Celestial equator: ————. Sun's path (ecliptic) ————.

London, therefore, the zodiac rises in its steepest position to $38\frac{1}{2}° + 23\frac{1}{2}° = 62°$, whereas in its lowest position it is only $38\frac{1}{2}° - 23\frac{1}{2}° = 15°$ above the horizon. The ecliptic's variation over the course of a day is therefore quite considerable. At the same time, the angles of 62° and 15° are the limits of the inclination of the zodiac to the horizon.

Together with the rising and sinking of the zodiac in the course of a day, there occurs a horizontal displacement of the *rising* and *setting* points along the horizon. The latter are always situated at opposite points on the horizon. At the highest and lowest positions of the zodiac they are exactly in the east and west; at the intermediate positions they are roughly south-east and north-west or north-east and south-west (Figures 13 and 15). This fact can be of help when we are looking for the zodiac in the sky. Once we have found one constellation near the horizon, we can seek its counterpart on the horizon opposite. The same principle of orientation can be applied to every visible constellation of the zodiac. If we imagine our line of vision extended backwards, it will invariably point to the opposite

33

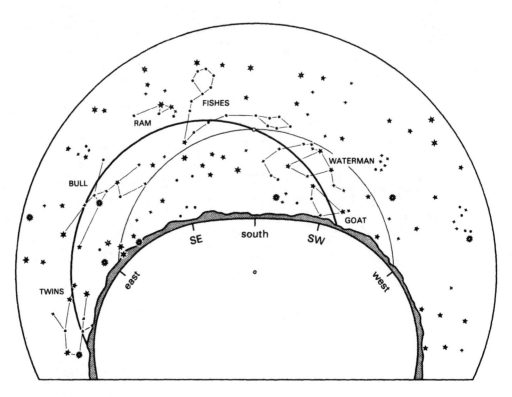

Figure 15. Intermediate situation of the zodiac, displaced to the east, as on March 21 at 12.00, June 21 at 6.00, Sep 23 at 24.00, Dec 22 at 18.00. Celestial equator: ———. *Sun's path (ecliptic)* ———.

constellation below the horizon. Thus, when we have located the Bull, we can easily find the direction to the Scorpion; similarly, the Lion gives us the direction of the Water-Bearer. In this way the visible portion of the zodiac shows the direction of the invisible portion below the horizon, allowing us to visualize its overall spatial situation at any moment.

A comprehensive diagram of the displacement of the zodiac in the course of a day, at intervals of two hours, is given in Figure 16. The zodiac's movement at its intersection with the horizon is comparable to that of a pair of scissors opening and closing while being moved backwards and forwards. For simplicity, we shall call it a 'scissoring' motion, as it appears in a whole series of astronomical phenomena which we shall return to later in connection with the Moon's orbit and the movements of the planets.

The positions of the zodiac at different times of the day and year can be deduced from the observation of the Sun's movements.

With the aid of a two-sided rotating star-map (see note on p. 28) the transitions

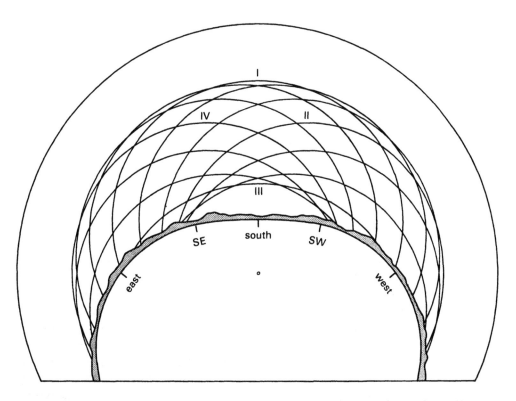

Figure 16. Positions of the ecliptic over the horizon at intervals of two hours. I, II, III, and IV correspond to Figures 12 to 15 respectively.

from one position to the next can be followed continuously, and the corresponding hours and dates of the year read off directly. On this star-map the visible heavens are portrayed on two separate faces, showing the view towards the north and the south, respectively. The scissoring daily movement of the zodiac over the southern horizon corresponds to the impression of direct observation.

A closer look at Figures 12 to 15 reveals another interesting phenomenon. Since the daily rotation of the celestial sphere is parallel to the celestial equator, the constellation of the Virgin rises steeply over the horizon; about three hours are needed before it has completely appeared. The Fishes, by contrast, stand at a much flatter angle to the horizon when rising, and require only about one hour before they are completely visible. In setting, the relationship is reversed (compare also Figures 45, 46 and 51). The other constellations of the zodiac occupy intermediate positions.

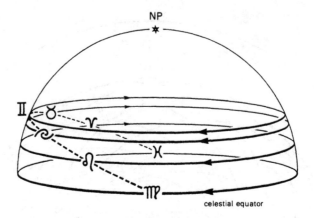

Figure 17. Position and movement of the zodiac at the Earth's North Pole.

Let us look briefly now at the corresponding situations at the Poles and at the equator, and then at the southern hemisphere.

At the *North Pole* the celestial equator coincides with the horizon, and the stars describe horizontal circles. From this it is clear that the vernal and autumnal equinoxes lie on the horizon, which means that only the upper half of the zodiac is visible; the Fishes and the Virgin are each only partly visible. The inclination of $23\frac{1}{2}°$ to the equator is especially striking, as the latter coincides with the horizon. The constellations of the Bull and the Twins describe a horizontal circle during the course of a day at a height of $23\frac{1}{2}°$; the Ram, the Crab and the Lion describe correspondingly lower circles; the Fishes and the Virgin wander along the horizon from left to right (Figure 17). At the South Pole it is the southern half of the zodiac which carries out the same movements, but in the opposite direction (right to left).

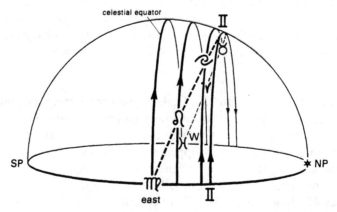

Figure 18. Position and movement of the northern part of the zodiac at the Earth's equator.

36

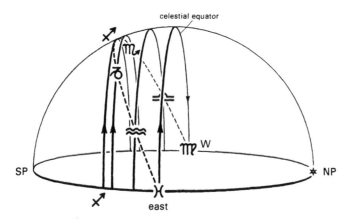

Figure 19. Position and movement of the southern part of the zodiac at the Earth's equator.

At the Earth's *equator* the two halves of the zodiac can be observed under equally favourable conditions. Here the celestial equator rises vertically from the east and west points of the horizon to the zenith. If the Fishes are in the west and the Virgin in the east, it is the northern half of the zodiac which arches over the landscape. The constellations of the Bull and the Twins are inclined at 23½° to the north of the zenith (Figure 18). When, after 12 hours, the Fishes and the Virgin have reversed their places, the southern arc of the zodiac stands over the Earth with an inclination of 23½° to the south of the zenith (Figure 19). The various intermediate positions can easily be imagined. At the equator, the zodiac swings within a belt of ± 23½° back and forth through the zenith. Similarly, the rising and setting points swing 23½° to the north and south of the east and west points.

In the *southern hemisphere* the high zodiacal constellations are those which are seen in the north as the low summer constellations. In the winter night sky, therefore, the Scorpion and the Archer culminate near the zenith, and in the summer night sky the Bull and the Twins describe relatively low arcs over the southern horizon. It should be borne in mind, however, that the visible constellations of the Scorpion and the Archer lie almost entirely to the south of the ecliptic. For this reason they describe even higher arcs in the sky of the southern hemisphere than the midsummer sun. This relationship can be seen in Figure 20, which shows the steepest position of the zodiac, as it appears for example in winter on June 21 at midnight. In Sydney and Cape Town (34° S), therefore, both constellations culminate in the zenith; in Brisbane and Johannesburg (27° S) the southernmost stars of the Scorpion extend about 15° beyond the zenith southwards. In London (51½° N), by contrast, the Hyades, the most salient stars of the Bull, culminate more than 30° south of the zenith, and the bright star Castor in the Twins, which represents the northernmost part of the zodiac, only approaches to about 20° of the zenith. In Florida and Texas (31° N), Castor just reaches the

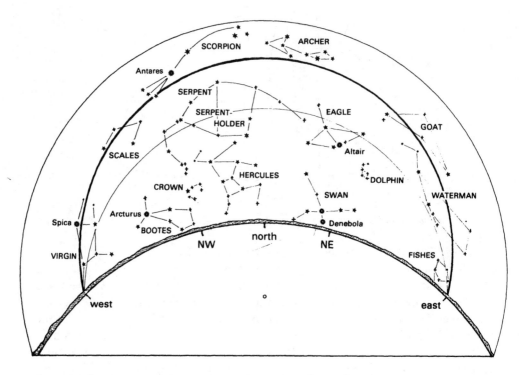

Figure 20. Steep position of the zodiac in the southern hemisphere, as on March 21 at 6.00, June 21 at 24.00, Sep 23 at 18.00, Dec 22 at 12.00. Celestial equator: ———. Sun's path (ecliptic) ———.

zenith, but even here the greater part of the northern constellations is inclined decidedly to the south. Even in Christchurch, New Zealand, the southernmost large city in the world (at $43\frac{1}{2}°$ S) the southernmost stars of Scorpion culminate in the zenith. Correspondingly, the summer constellations of the southern hemisphere rise somewhat higher in the sky than is the case in the northern temperate zones. This position is illustrated in Figure 8.

The Sun

4. The Sun's daily and yearly course

If in our consideration of the celestial phenomena we leave the Sun out of account, as we have done in describing the daily movements of the stars, the constellations and the zodiac, we are left with nothing more than the rather monotonous twenty-four-hour daily rotation of the entire sky.

The great life-rhythms, the alternation of day and night and the course of the seasons, are gifts of the Sun to the Earth. With respect to the celestial phenomena as well, we encounter variation, metamorphosis and the rhythmic alternation between polar extremes as soon as we follow the Sun in its positions and movements. For our immediate experience the Sun, like the Moon and the planets, is a 'wandering star'. In contrast to the fixed stars, which retain their relative positions almost unchanged, thereby maintaining the forms of the constellations for millennia, the wandering stars constantly change their positions on the celestial sphere. They do, however, remain within a certain zone: the girdle of the twelve constellations of the zodiac.

The phenomena connected with the daily and annual course of the Sun are, in greater or lesser detail, familiar to everyone. And yet it is difficult to maintain full consciousness of the movements which take place over and around us day by day, and to bring them together into a clear and versatile conception of the whole. Such a comprehensive view can only be reached in stages.

On the other hand, the observation of the Sun's daily and yearly movements, which in itself is easy to carry out, facilitates an understanding of the stars as well, for example in looking for the constellations of the zodiac.

The daily movement of the Sun

Figure 21 illustrates the daily movement of the Sun at different times of the year, looking south. It will be seen that all the diurnal arcs of the Sun lie symmetrically to either side of the meridian, the line from south point through zenith. The steeper the Sun's culmination, the greater is the span of its arc over the landscape, and the further the rising and setting points are displaced to the north-east and north-west. At a latitude of 50° the latter wander over an area of 76°, covering almost a quarter of the horizon.

The rising and sinking of the diurnal arcs takes place in a rhythmic pendulation, like a sine wave. Figure 21 shows this rising and sinking for twelve equal intervals within the year. Starting from the middle position, the celestial equator, it swings 23½° in either direction to the extreme diurnal arcs of the summer solstice on June 21 and the winter solstice on December 22. On these days the Sun attains its highest and lowest arcs, respectively. When the Sun stands near the celestial equator in spring and autumn its declination changes most rapidly. In the twenty-

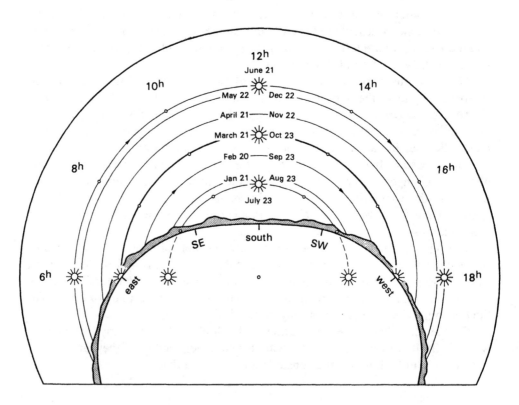

Figure 21. The rising and sinking of the diurnal arcs of the Sun during twelve equal intervals in the year. Celestial equator: ———.

40

4. THE SUN'S DAILY AND YEARLY COURSE

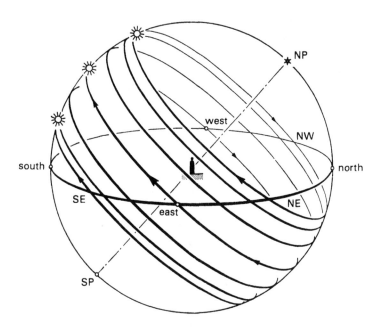

Figure 22. Spherical representation of the diurnal arcs of the Sun during twelve equal intervals in the year (as in Figure 21).

six days from March 21 to April 16 it rises the first 10° from the equator (declination 0°); for the next 10°, to a northern declination of 20°, the longer period of thirty-four days is required, from April 16 to May 20. By this time, however, the rising movement has decelerated so strongly that for the last $3\frac{1}{2}°$ to the summer solstice on June 21 another thirty-two days are needed. A corresponding relationship holds true for the other three quarters of the year.

Figure 21 shows the daily arcs divided into segments of two hours. Six hours before true noon* at the summer solstice, for example, the Sun has already been above the horizon for some time. At the times of equinox, on March 21 and September 23, when the Sun's path coincides with the celestial equator, it is, at the same hour, only just rising in the east; at its southernmost declination in the winter it is still below the horizon.

Any given diurnal arc of the Sun can only be approximated by a circle, although the deviation is very slight. For the 365 daily arcs pass over, one into the next, in a continuous spiralling movement.

Figure 22 portrays the different daily arcs which the Sun traces above and below the horizon. Looking from the centre towards the south, we see the arcs

---

*In Chapter 6 we shall discuss the difference between 'true noon' and the mechanical regularity of 'clock noon'.

characteristic for the different times of year. If we imagine the path as a single continuous line, we would have a comprehensive picture of the Sun's path in the course of a year as a spherical spiral which, in 365 revolutions, covers a girdle of the heavens whose breadth is twice $23\frac{1}{2}°$.

The conceptions which we have so far developed may all be derived from a study of the Sun's changing position with respect to the horizon in the course of a year.

The annual movement of the Sun

A quite new aspect arises when we inquire into the Sun's relationship to the fixed stars. This cannot of course, be perceived directly, for the Sun's light renders the stars invisible by day. It can, however, be deduced.

During the night all the fixed stars and constellations in the southern sky describe arcs similar to those described by the Sun in the daytime. In contrast to the Sun, however, they maintain the same arcs from day to day; in other words, the fixed stars undergo no annual change in declination. It follows from this that the Sun must change its position relative to the fixed stars in the course of the year. This is immediately apparent from the changing countenance of the starry heavens from week to week at the same hour of night (compare Chapter 2).

The autumn and winter sky is characterized by the brilliant constellations of Orion, the Great Dog (Canis Major) with Sirius, the Charioteer (Auriga), the Bull and the Twins, whereas the night sky of summer is marked by the Scorpion, the Archer, and above all by the Lyre (Lyra), the Swan (Cygnus) and the Eagle (Aquila). For the southern hemisphere the opposite seasons must be taken: Orion is visible in Spring and Summer, the Swan in winter.

The annual rhythm of the appearance and disappearance of the constellations reflects the progression of the Sun in its course through the fixed stars, which cannot in itself be seen. The Pleiades, for example, are still visible in April above the point of sunset; within a few weeks they have disappeared, to emerge again at the beginning of July in the early morning sky. The Sun has therefore moved from a position to the west (right) of the Pleiades, where it set before the constellation, to a position to the east (left) of it. Stars visible in the evening sky gradually fade into the twilight to appear again, after a period of invisibility, in the morning sky.

The west-to-east component of the Sun's movement brings about an increase in its right ascension (AR). As we have already mentioned, this displacement is reflected in the fact that all the individual stars and constellations rise earlier from night to night. Any given star will rise $3^m 56^s$ earlier each night (compare Chapter 2). This corresponds to the Sun's daily movement of 1° (roughly twice the diameter of the Sun's disc) towards the east, against the background of the fixed stars. A sidereal day is therefore shorter than a solar day.

The daily discrepancies ($3^m 56^s$) between the sidereal and solar days total about

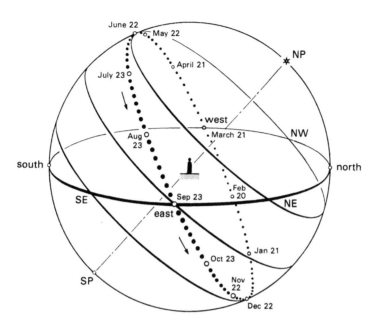

Figure 23. The inclined path of the Sun (ecliptic) in its annual course seen in relation to the diurnal arcs of movement.

two hours in a month; in half a year twelve hours, and in a year a day. In a year of 365 solar days, therefore, the fixed stars rise and set 366 times.

Let us now bring together the two aspects of the Sun's movement, which we have thus far regarded separately. The Sun wanders, in its consecutive positions, once each year through the whole of the fixed star sphere, which is now to be imagined at rest (Figure 23). It advances along a great circle which crosses the daily circles of movement at an angle of 23½°, the obliquity of the ecliptic. The twelve constellations through which it passes are the zodiacal constellations. This annual path, or ecliptic, is illustrated in Figure 23 by consecutive positions of the Sun at intervals of roughly 4 days.

From month to month the Sun advances from one constellation of the zodiac to the next and displays, through the position and height of its daily arcs, its rising and setting points on the horizon and the relative length of day and night, the features which are characteristic for each of these constellations. The Sun's course through the zodiac, and its positions on the first and fifteenth of each month are shown on Figure 24.

In the previous chapters we have described how the ecliptic and zodiac change their position with respect to the horizon, and thereby alternate between high and

43

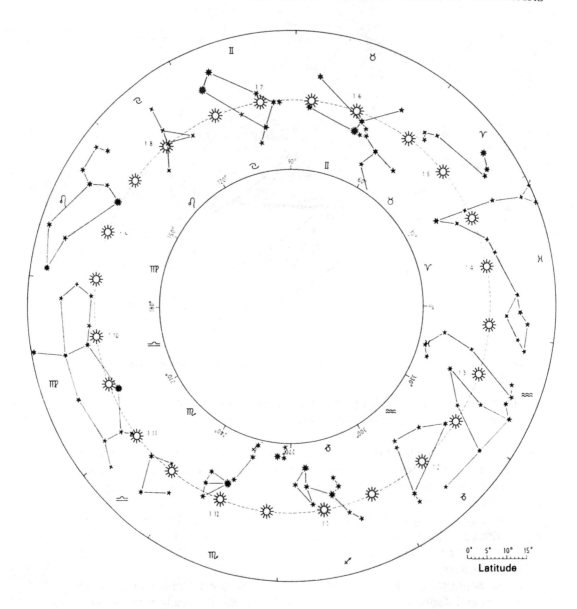

Figure 24. The Sun's orbit in the zodiac (ecliptic), with the Sun's position on the first and fifteenth of each month. On the outer edge the boundaries of the zodiacal constellations are indicated; on the inner edge the regions of the zodiacal signs are shown with their boundaries in ecliptical longitude. The displacement of the signs from the constellations due to the precession of the vernal equinox (0°) can be seen clearly.

low arcs. According to the Sun's location in the zodiac, however, the different arcs of the zodiac over the horizon (Figures 12–15) occur at different times of day.

The steepest position (Figure 12) occurs at the

Summer solstice (June 21)	at noon	Sun in the Twins
Autumnal equinox (Sep 23)	at 6^h	Sun in the Virgin
Winter solstice (Dec 22)	at midnight	Sun in the Archer
Spring equinox (March 21)	at 18^h	Sun in the Fishes

The lowest position (Figure 14) occurs at the

Winter solstice (Dec 22)	at noon	Sun in the Archer
Spring equinox (March 21)	at 6^h	Sun in the Fishes
Summer solstice (June 21)	at midnight	Sun in the Twins
Autumnal equinox (Sep 23)	at 18^h	Sun in the Virgin

The occurrences of the intermediate positions (Figures 13 and 15) can be deduced from this table.

In the southern hemisphere the steepest position occurs at the times indicated in the second table (lowest position in the northern hemisphere), while the lowest position corresponds to the times in the first table (steepest position).

Such observations make clear the relationship between the course of the day and the course of the year. Were it possible to follow the night sky uninterruptedly for 24 hours, the observer would see the same changes recapitulated, which he would otherwise perceive over the course of a year by observing the positions of the stars at the same hour each night, for example at midnight.

The daily movement of the Sun in its changing positions can be followed with relative ease on a moveable star-map.

It will be clear from the foregoing that the Sun wanders through the fixed star sphere in the course of each year, following a path projected onto the zodiac, which we call the ecliptic. As a result of this annual movement, the Sun appears to be falling continually behind the 24-hour daily movement of the heavens. Whereas the latter proceeds from east to west, the annual movement of the Sun through the zodiac takes place in the opposite direction, from west to east (Figure 23).

The orbital plane of the ecliptic is not only defined by the movement of the Sun. With slight deviations the Moon and the planets also move along this path. Strictly speaking, it could be regarded as an equator or 'equalizer' of their movements, whereas the celestial equator maintains its equalizing function with respect to the daily movement of the fixed stars. For this reason, a second co-ordinate system has been established with the ecliptic as its equator, its northern pole in the constellation of the Dragon (AR 18^h, δ +66° 33′) and its southern pole in Dorado near the large Magellanic Cloud (AR 6^h, δ –66° 33′). Ecliptical longitude λ is measured in degrees from 0° to 360°, beginning with the point of vernal equinox, and proceeding from west to east, in the direction of the sun's movement. The distance to the north or south of the ecliptic is defined as latitude β, and counted in degrees from ± 0° to 90°. Thus, any point in the sky can be localized in either co-ordinate system.

Generally speaking, astronomical handbooks make use of the equatorial system and give locations in right ascension and declination (AR, ± δ), while astrological ephemerides give the ecliptical co-ordinates (λ and β). The reason for this can be understood out of the different requirements of astronomers and astrologers. Astronomical telescopes are mounted on an axis parallel to the earth's rotational axis. Hence, the declination of a star or planet can easily be fixed, so that the telescope will move only along the chosen circle of declination, thus following its arc of daily movement. This would not work for the ecliptical co-ordinates, which are not related to the daily movement. The astrologer, on the other hand, wishes to know the relationship of the planets to their central plane of movement, the ecliptic. It follows that all geometrical relationships or *aspects* of the planets, defined as conjunctions, oppositions, and so on, can be determined with reference to either system. Some of the consequences of this discrepancy are described in Chapter 29.

The southern hemisphere

Strictly speaking, the solstices and equinoxes are at the opposite points in the southern hemisphere, as the seasons are at the opposite time of year. The vernal equinox, for example, is in the constellation of the Virgin, and is crossed by the Sun in its annual movement on September 23. As reference cannot always be made to this in the present book, readers in the southern hemisphere must pause in each case to consider what is meant. This applies equally to the temporal and spatial relationships which are described for the northern hemisphere. The following rules of thumb may be of help:

Remain unchanged:	*Sometimes changes:*	*Always changes:*
celestial east/west	summer/winter	celestial left/right
terrestrial east/west	spring/autumn	celestial above/below
celestial north/south	(as the sense requires)	terrestrial north/south (when referring to celestial phenomena)

Most diagrams showing parts of the sky, planets, Sun, Moon and so on, must be looked at upside down.

Apogee and perigee of the Sun

Exact observation shows that the Sun does not move at a constant, but at a rhythmically changing pace through the zodiac. The mean daily movement along the ecliptic of almost 1° (more exactly, 59′ 8″) becomes somewhat slower during the summer half-year (minimum 57′ 10″ on July 4) and somewhat faster in the winter half-year (maximum 61' 9" on January 2). However slight the difference, its total effect produces a noticeable asymmetry in the course of the year. The Sun passes through the upper half of the ecliptic, from the vernal to the autumnal equinox, somewhat more slowly than through the lower half. The summer half-year in the northern hemisphere (from March 21 to September 23) with $186\frac{1}{4}$ days is therefore 8 days longer than the winter half-year of only $178\frac{3}{4}$ days. The four seasons are of unequal length.

As early as 150 BC the Greek astronomer Hipparchus was aware of this discrepancy. Using the simplest of instruments, a shadow-staff sunk vertically into the ground, and a shadow-casting ring mounted in the plane of the equator, he was able to determine the days of solstice and equinox, thereby discovering the unequal length of the four quarters of the year.

Parallel to the unequal velocity of the Sun, telescopic observation shows that the Sun's disc, whose diameter measures about $\frac{1}{2}°$, is not constant in size, but is smallest at the time of its slowest zodiacal movement on July 4, and largest on January 2 (see Appendix, Table 2.1). From this it can be concluded that the Sun's distance from the Earth is also subject to a yearly cycle: on July 4 it stands at its farthest point from the Earth (*apogee*), at which its diameter is smaller than at its nearest point (*perigee*) on January 2. (Conversely, we may speak of the aphelion and the perihelion of the Earth).

These phenomena lead to the concept of an eccentric solar orbit around the Earth. (As further investigation shows, this orbit can be thought of as a nearly circular ellipse.)

5. The Sun's course through the zodiac

Figure 24 shows the girdle of the zodiac and the Sun's orbit. The connection with Figure 23 may easily be found. The portion of the ecliptic which stands above the horizon corresponds to the upper half of the star-map. In both cases the ecliptic is shown in its steepest position, which it occupies, for example, at the end of June at midday, or at Christmas at midnight.

From the *constellations* of the zodiac can be determined which stars are actually in the immediate vicinity of the Sun on any given day of the year. Because of the asymmetry and unequal size of the individual constellations, the Sun requires different intervals of time to pass through them. The days on which the Sun passes into the various constellations are given in Table 2.4 in the Appendix.

The traditional zodiac of twelve constellations as it appears on most modern star maps is a heritage of ancient folk tradition. Although the majority of constellations in the sky as a whole are Greek, most of the zodiacal constellations seem to have their origin in Babylonian mythology. It is likely that they were adopted and to some extent integrated into Greek mythology.

The divisions between the single constellations may have varied considerably at different times and in different parts of the classical world. The divisions described by Ptolemy in his *Almagest* and handed down to modern times, should not therefore be seen as anything final and ultimate. Indeed, the strikingly different sizes of some of the constellations would seem to point to rural folk-traditions rather than to mystery teachings. One can hardly avoid having the impression that the divisions were arrived at in a visual, pictorial way, without too much concern for the symmetry of celestial architecture. These consellations belonged to the lore of peasants and shepherds, who ordered their labours according to a living calender of the sky. The seasonal appearances and disappearances of the constel-lations and their connection to work on the land were described by Hesiod in his epic poem *Works and Days*.

This pictorial, folk zodiac of visible constellations is in stark contrast to the zodiac of *signs*, which we owe to *Hipparchus* (*c.* 150 BC). In Hipparchus' system the Sun's orbit is divided into equal segments, beginning from the vernal equinox. The segments are designated with the Latin names of the constellations in their proper sequence. The spring quarter is thus distinguished by the signs Aries ♈, Taurus ♉, Gemini ♊; from the time of the summer solstice there follow the three 'summer' signs Cancer ♋, Leo ♌, Virgo ♍. Following the autumnal equinox come the further divisions Libra ♎, Scorpio ♏, Sagittarius ♐, and after the winter solstice the Sun wanders through the 'winter' signs Capricornus ♑, Aquarius ♒, Pisces ♓.

At the time when it was established, that is, about two thousand years ago, this classification corresponded to the constellations of the starry heavens. The vernal equinox lay on the boundary between the constellations of the Fishes and the Ram; it could, in agreement with the visible constellation, be called the First Point of Aries. When the Sun crossed this it passed simultaneously into the 'sign' and the 'constellation' of the Ram. And, with the symmetrical distribution of the constellations, the signs and constellations corresponded fairly well in other parts of the zodiac as well at that time.

Since that time, the precession of the equinoxes (see Chapter 10) has brought about a marked separation of signs and constellations. Since the vernal equinox regresses (moves clockwise along the ecliptic) by 30° in roughly 2160 years, it has now moved from the Ram to well past the middle of the Fishes. Thus the old division into signs already deviates by a full constellation from the original positions. The difference in question can be seen on Figure 24 where the beginning points of the signs are indicated on the inner edge of the picture.

The Greek division of the zodiac into signs has been handed down to the present age through two quite different channels which have, since ancient times, proceeded in opposed directions: on the one hand through the astronomical, and

48

on the other hand through the astrological tradition of the Middle Ages. In the realm of spherical astronomy this division serves in a logical way as the foundation for a mathematical, numerical subdivision of the ecliptic in longitudinal degrees. The inherited symbols were retained, so that the sign ♈ is used for the starting-point at the vernal equinox; similarly, the symbol ♎ is used for the autumnal equinox, and signs ♋ and ♑ for the summer and winter solstices.

Starting from totally different premises, and with quite different aims in mind, the astrological tradition also makes use of this division. It attributes to the zodiacal *signs* dynamic qualities and inner significance with respect to the Sun's movement. Many astrologers, even today, continue to dismiss the circle of constellations as meaningless, and direct their attention only to the signs; indeed, many seem quite oblivious of the very existence of the discrepancy.

An unprejudiced view of the differences between the two systems need not lead to the exclusion of the one or the other, but rather to an understanding of where the different standpoints can lead. Thus we can discover that two essentially different components interplay in the Sun's yearly rhythm.

The one emerges when we direct our attention solely to the interrelationship between Sun and Earth in the course of the year. The mutual change of position leads, in the course of months, to the rhythmic pendulation of the seasonal path of the Sun for the different parts of the Earth, which largely determines the annual life-cycles on our planet.

The other component has to do with the Sun's changing position in the fixed star sphere. Expressed in astronomical concepts, the passage of the Sun through the circle of constellations brings about a natural division of the sidereal year (the journey of the Sun from a certain fixed star to the same fixed star again).

The division of the year's course into signs underlies the twelvefold structure of the tropical solar year. As the term 'tropical' implies (from the Greek τροπή, turning) this rhythm is related solely to the periodical return of the Sun's 'turning points', or solstices, that is, the yearly alternation of its highest and lowest culminations in summer and winter. No account is taken of its position with respect to the fixed stars. The tropical year is thus directly connected with the annual breathing rhythm of the Earth, which can be followed outwardly in the dynamic rising and sinking of the Sun's culmination point, and which, as a life-rhythm of our planet, is at work in the formation and development of all living creatures, as well as in numerous processes of inorganic nature.

In contrast to the tropical year, the sidereal revolution of the Sun through the constellations of the zodiac can be described as a circulating movement, as the Sun's journey through the various spatial directions of the cosmos.

A further possibility of dividing the zodiac is one prevalent in Babylon from about the seventh century BC onwards. This was a division into twelve equal segments, but based on the constellations, rather than the tropical points. In this system the bright star Aldebaran in the Bull was taken as the centre of that

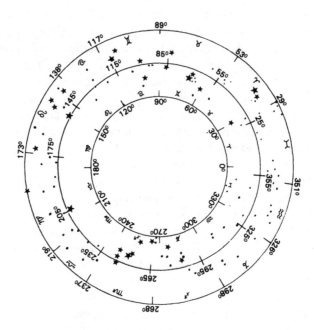

Figure 25. The Babylonian zodiac divided into twelve equal segments based on the constellations (middle circle). The outer circle shows the constellations and their division into unequal segments; the inner circle is the regular division into signs.

constellaton, and the starting point for the 30° segments, which thus gave the midpoint of each constellation. It is especially striking in this system that the two brightest stars in the zodiac, Aldebaran in the Bull and Antares in the Scorpion, stand almost exactly opposite each other in the sky and form the mid-points of their respective constellations. This system strikes a kind of balance between the requirements of the thinker and the observer. Just as the thinker demanded equal and symmetrical divisions of the sky as a sound basis for all mathematical work, the observer wished these divisions to be related to the visible stars.

Figure 25 illustrates the three zodiacs we have described in their relation to the fixed stars. The central circle, which is the ecliptic, is divided into the equal segments of the Babylonian zodiac. The outer circle represents the divisions of the traditional zodiac of unequal constellations, and the inner circle represents the tropical signs. The divisions of the central and outer circles may be thought of as permanent, while those of the signs slowly wander in a clockwise direction as a result of the precession of the equinoxes.

For the purposes of this book, all illustrations show the forms of the constellations according to the traditional Greek system. They are always designated with the English names: Ram, Bull, Twins, and so on; while the astrological signs are designated in Latin: Aries, Taurus, Gemini, and so on.

50

6. The geography of sunrise and sunset

The times of sunrise and sunset fluctuate during the course of the seasons. Parallel to this, the points of rising and setting wander from day to day along the eastern and western horizons. They do not do so at an even rate, but follow a rhythmic pendulation. The greatest change from one day to the next, both in time and in space, occurs in spring and autumn. At the times of the summer and winter solstices the differences are very slight.

To see how differently a sunrise or sunset is orientated in spring or autumn, in summer or winter, we may attempt to answer the question: from which points on the Earth can a single, individual sunrise or sunset be seen simultaneously?

The zone of morning twilight, which separates the regions of day and night over the Earth's surface, naturally extends far beyond the visible horizon of an individual observer. It leads into other regions and lands, which experience the break of day and the sunrise at the same moment. Such lines of simultaneous sunrise lie in quite different directions over the Earth's surface in different seasons. These directions may easily be found: if we face the point of sunrise exactly, and stretch out our arms to either side, they will point in the direction which we are seeking.

At the times of vernal and autumnal equinox, when the Sun rises exactly in the east, the morning twilight zone correspondingly runs exactly north-south, coinciding with the geographical meridian. Since, in the course of each day, the sunlight travels from eastern to western lands, the line of sunrise wanders over the Earth on March 21 and September 23 as portrayed in Figure 26. The individual lines illustrate the progression of the sunrise from east to west at one-hour intervals. It can be seen, for example, that sunrise takes place simultaneously in Sweden, Poland, Hungary, Albania, Libya, Angola, and the Cape of Good Hope.

A corresponding observation may be carried out for the sunset. Since the Sun sets exactly in the west on March 21 and September 23, the line of simultaneous sunset also wanders on these days over the Earth as shown in Figure 26.

Quite different is the course taken by the boundary between day and night on June 21 and December 22 (Figures 27 and 28). The points of sunrise move towards the north-east in summer and the south-east in winter. At the times of the solstices, therefore, the lines of simultaneous rising and setting stand obliquely to the geographical meridians. In summer the lines of simultaneous rising run from the north-west to the south-east, for example, from Greenland through the British Isles, France and Sicily, Eastern Africa and Madagascar. The lines of setting, by contrast, proceed from the north-east to the south-west. In winter the situation is

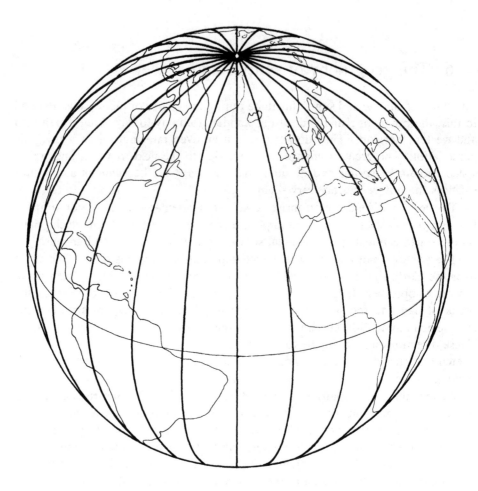

Figure 26. Lines of simultaneous sunrise and sunset on March 21 and Sep 23 (Sun in Fishes and Virgin respectively). The lines are at one hour intervals.

reversed: for example, simultaneous rising in the northern Urals, Leningrad, northern Poland, Switzerland, eastern Spain, the bulge of West Africa and the Atlantic, reaching the Antarctic Circle west of South America. At other times of year the lines assume an intermediate angle. In this way the lines of simultaneous sunrise and sunset connect the locality of the observer with ever-changing regions of the Earth in the course of a year.

The various tilts of the rising and setting lines in summer and winter reflect the annual movement of the Sun to northern and southern declinations. In the summer, when the Sun is in northern declination, the northern regions of the Earth receive more light. At this time of year a certain region around the North Pole remains constantly within the field of the Sun's light, enjoying uninterrupted

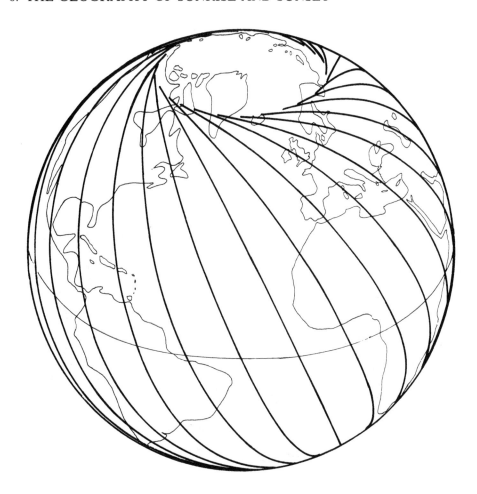

Figure 27. Lines of simultaneous sunrise on June 21, or sunset on Dec 22 (Sun rising in the Twins, setting in the Archer). The lines are at one hour intervals.

day. On June 21, when the Sun stands at a northern declination of $23\frac{1}{2}°$, this region extends $23\frac{1}{2}°$ to the south of the North Pole. The boundary, where once every year, at the summer solstice, the Sun does not set, lies at $66\frac{1}{2}°$ northern latitude, and is known as the *Arctic Circle*. The line of sunrise at the summer solstice, therefore, glides along the Arctic Circle (Figure 27), and, in its daily movement, encloses the continually illuminated region of the Earth around the North Pole.

As the year progresses, the area of uninterrupted day becomes smaller; by the beginning of autumn it has shrunk together to a point at the Pole. From the time of the summer solstice until the autumnal equinox, the twilight zone glides along circles of latitude which approach nearer and nearer to the Pole. At the same time the lines approach closer and closer to the north-south direction, which is attained

53

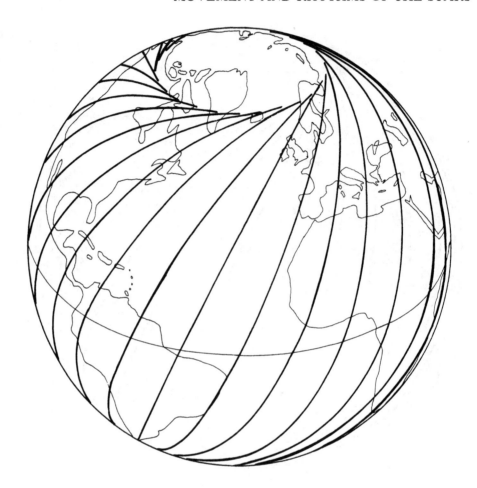

Figure 28. Lines of simultaneous sunrise on Dec 22, or sunset on June 21 (Sun rising in the Archer, setting in the Twins). The lines are at one hour intervals.

on September 23. From this time onwards there is a region of constant night around the North Pole which reaches its greatest expansion to the Arctic Circle on December 22. Along its boundary glides the zone of constant twilight (Figure 28). The Sun then stands in southern declination and draws the region around the South Pole into its constant field of light. Thus an area of continual day and continual night spreads over the polar regions in half-yearly alternation. At the time of the solstices they attain their greatest extension as far as the Arctic and Antartic Circles, while at the equinoxes the reversal takes place at the points of the Poles. It is an alternating expansion and contraction of a zone of continual daytime and one of continual night-time: polar day and polar night. Between these two areas lies the region in which day and night alternate. The latter, however, is not uniform. Going out from the equator, where the length of day and night is always equal, towards the region of polar darkness, the night becomes longer and longer. Proceeding in the opposite direction, towards the region of polar light, the day becomes longer.

The movements of the lines of sunrise and sunset can be called 'scissoring' (see page 34). They do not remain parallel but slowly and continuously pass over into the various slanted positions of the different seasons. This movement is indicated in Figure 29. The horizontal line represents a given circle of latitude; the various oblique lines show the changing slant of a line of simultaneous sunrise or sunset during the course of a year. It can be seen that, on either side of a line of latitude, the Sun's light field alternately moves slightly ahead of, or falls slightly behind, the regular daily movement of the Earth. Out of the interplay between the even daily rotation of the Earth and the seasonal change in obliquity of the lines of sunrise and sunset arises the 'scissoring' movement of the latter around their northern boundary.

The obliquity of the twilight zone increases as we move from the equator towards the Poles. On the equator the maximum deviation from the meridian is ± 23½°; on the polar circles it is ± 90°. Figure 30 illustrates this principle by showing the extreme deviations of the rising and setting lines in both directions.

We now have a living picture of how every locality on the Earth experiences the course of the year differently, and how the Earth's fields of light and darkness move through the year in their dynamic play of movement.

The scissoring movement thus acquires a quite real significance for the whole surface of the Earth. For in the areas of the Poles, the polar circles and the equator, and above all in the regions of morning and evening twilight, numerous processes of nature, both organic and inorganic, are dependent on it.

Because of the scissoring movement of the lines of rising and setting, tables of sunrise and sunset are only valid for one particular place on the Earth. For other places they must always be recalculated. In the case of localities on the same geographical latitude, the difference in the times of sunrise and sunset remains

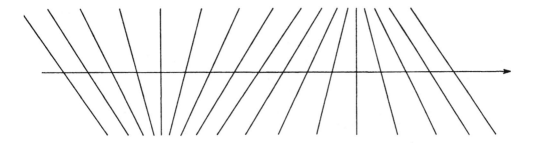

Figure 29. The 'scissoring' movement of the lines of sunset and sunrise.

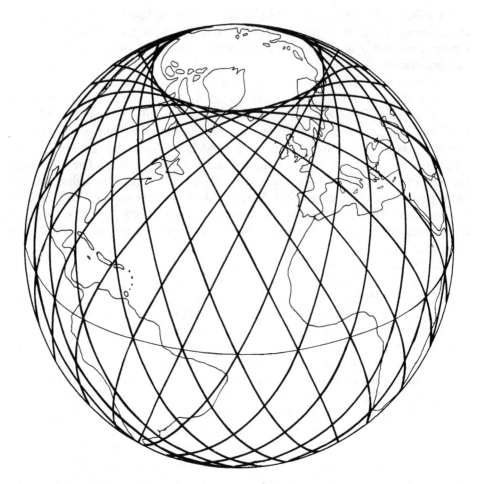

Figure 30. The lines of simultaneous sunrise and sunset at the time of the solstices (compiled from Figures 27 and 28).

constant during the whole year. It can easily be derived from the difference in longitude of the two places in question.

The line of sunrise wanders daily around the Earth from east to west at 360° in 24 hours, which is 15° in 1 hour. Knowing the longitude of two cities on the same latitude (for example Naples, 14°E, 41° N, and New York, 74° W, 41° N) we can easily calculate that the difference in longitude of 88° causes a time difference of nearly 6 hours. Sunrise and sunset therefore both take place 6 hours earlier in Naples than in New York.

The situation is quite different when we compare localities with different latitudes, as for example Naples and London (0°, 51½° N). Only for March 21 and September 23, when the rising and setting lines run parallel to the meridians, can the time difference in sunrise and sunset be derived directly from the difference in longitude, in this case 14°. On the days of the equinoxes, therefore, the Sun rises and sets almost an hour earlier in Naples than in London. At other times of the year, when the scissoring displacement of the rising and setting lines comes to expression, this is not the case. As Figures 27 and 28 show, the sunrise on June 21 takes place roughly simultaneously in both cities. On December 22, on the other hand it takes place more than $1\frac{1}{2}$ hours later in London than in Naples. We can see that the difference in the time of sunrise for two cities is not constant, but subject to an annual fluctuation. Only when the cities both lie on the same latitude does the fluctuation not occur. The times of sunrise and sunset may be read off graphically for localities in a whole region if the lines of simultaneous sunrise and sunset are charted on a number of successive maps.

These relationships may serve as a stimulus to become more conscious of the changing way in which the daily risings and settings in one locality relate to other lands and regions of the Earth.

Since every position of these lines corresponds to a certain seasonal position of the Sun in the zodiac, they can also be thought of as the connecting lines of all localities at which a certain sign of the zodiac simultaneously rises or sets.

The same basic phenomena and laws hold true for the rising and setting of the Moon. Here, however, a complete cycle of the scissoring movement of the rising and setting lines takes place in a sidereal month (27^d 8^h), that is, in the course of one full revolution of the Moon through the zodiacal constellations. If the additional deviations within the 18.6 year Moon-period are disregarded, Figures 26 to 30 can also be related directly to the positions of the rising and setting lines of the Moon, and the latter's positions in the zodiac. Figure 26 thus shows the situation when the Moon rises or sets in the constellations of the Fishes and the Virgin; Figure 27 is valid when the Moon rises in Bull/Twins or sets in Scorpion/Archer; and Figure 28 when it rises in Scorpion/Archer or sets in Bull/Twins.

The planets, too, undergo the same alternation in their rising and setting phenomena. The periods involved are their times of revolution through the zodiac, for instance, 12 years for Jupiter, 30 years for Saturn, and so on.

An overall picture of these phenomena would be obtained if, for a particular date of the year, the rising and setting lines of all the planets were determined and charted together. The various hemispherical light-fields of Sun, Moon and planets are simultaneously present over the surface of the Earth. But their relative positions and degree of interpenetration are constantly changing. They come together or draw apart from one another according to the changing positions of the respective luminaries.

7. The three suns and the equation of time

Our experience of time is not uniform, but is modified and differentiated by the daily and annual movements of the Sun. Astronomical indications are also brought into relation to the solar day and the solar year, and the duration of longer epochs of time is expressed in solar years. For qualitative experience, these seem to be obvious units. A quantitatively exact measurement, however, encounters certain difficulties.

The fact emerges, for example, that the solar day is not of constant length. Its true course may be followed by means of the shadow cast by the Sun on a sundial; not, however, through the uniform time-reckoning of a clock.

The designation 'day' is the result of the interplay between two factors, and can be defined as their mutual relationship;* it cannot, therefore, be identified with the diurnal rotation of the Earth alone. Following the celestial phenomena, we have various possibilities for determining the length of the day. We may, for example, take as our criterion the culmination of the Sun or of a certain fixed star. In this case the two factors are the horizon and the chosen luminary, resulting in either a solar day or a sidereal day. The two differ by about 4 minutes, as described in Chapter 2.

The length of successive sidereal days remains essentially unchanged, and is therefore suited to clock-measurement (sidereal time). This is not quite the case for the solar day. Our ordinary clocks indicate an average solar day, which deviates slightly from the true solar-day. Practically speaking, the deviation is inconsequential, at most ± 28 seconds per day. The difference between any two true solar days can therefore never be more than one minute. In the course of weeks, however, these slight differences add up to noticeable sums. These are calculated in minutes and seconds for every day of the year, and are known as the *equation of time*.† (All the equalizing values of astronomy, whereby true, irregular movements are led back to theoretical regular movements were, in earlier times, given the name of 'equations'.) The equation of time (e) is defined as the difference between the true solar time, or sundial time (s) and the mean solar time, or clock time (m): $e = s - m$.‡

---

*The same is true of all the rhythms of astronomy.

†See Table 2.3 of the Appendix and a solar ephemeris.

‡Until 1937 the starting-point was taken as the true solar time (s) to which the deviation (e) was added, giving the mean solar time (m): $s + e = m$. The new definition is a step to increasing abstraction, as a theoretical value is taken as the starting-point.

58

7. THE THREE SUNS AND THE EQUATION OF TIME

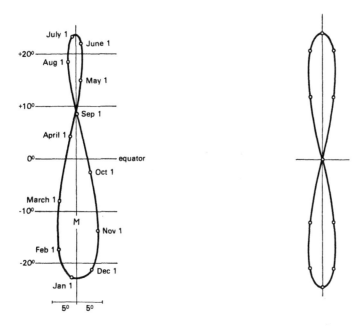

Figure 31. The analemma and, for comparison, a symmetrical lemniscate.

The equation of time can be shown graphically by charting the exact positions of the Sun on successive days at 12ʰ local time.* This gives the curve of the equation of time, or analemma, as illustrated in Figure 31. It shows not only the annual variation in the Sun's height of culmination, but also a horizontal deviation to the right and left (that is, west and east) of the meridian. These correspond to the temporal deviations resulting from the equation of time. The true Sun does not always reach its culmination (that is, the true solar noon) at the time of mean noon. As can be seen from the lemniscate, the Sun still stands east of the meridian at mean noon in February and July, while in May and November it has already crossed the meridian and stands to the west of it. In the course of the year, therefore, there are two maxima for the deviation towards the east and the west. On February 12 at mean noon the Sun stands 14½ minutes before its culmination. Expressed spatially, it stands about 3½° or 7 Sun-diameters from the meridian. On July 26 it stands nearly 6½ minutes (or 1½°) before culmination; on May 14 it has already crossed the meridian by 3¾ minutes (about 1°); and on November 3 the western maximum is attained of 16½ minutes (about 4°).

For our subdivision of the day, which depends on clocktime, this displacement results in the morning or afternoon becoming longer or shorter. In February, for

---

*Our clocks run on zone time, for instance Greenwich Mean Time (GMT) in Britain, or Eastern Standard Time (EST) in the eastern United States.

example, the mornings (that is, the time between sunrise and mean noon) are about half an hour shorter than the afternoons (the time from clock-noon to sunset). Conversely, at the beginning of November the afternoons are about half an hour shorter than the mornings. Four times each year, on April 16, June 14, September 1 and December 25 the value of the equation of time is zero. On these days the mean and true noon coincide; here the curve of the equation of time crosses the meridian.

The function and curve-form of the equation of time become evident in their structure when we consider how they arise out of the concepts of the three astronomical Suns. The following must be distinguished:

1. a mean Sun on the equator   moving with regular speed
2. a mean Sun on the ecliptic   ,,
3. the true Sun on the ecliptic   moving with variable speed

Our clock time, which counts average solar days and hours, only takes into account the *first* of these three Suns. The mean day would reflect the true movement of the Sun only if the latter's annual orbit were regular and lay on the celestial equator instead of the ecliptic.

The *second* Sun, the 'mean Sun on the ecliptic', is considerably nearer to the reality. It takes into account the Sun's annual course through the zodiac and its higher or lower declinations at midday. The noon positions of the Sun, however, vary slightly to either side of the meridian, for the two components of movement, which for the first Sun both lie on the celestial equator, now fall on to two great circles: the equator and the ecliptic. Their interplay is different from day to day. In other words: the daily increase in the Sun's right ascension is smaller at the time of the equinoxes (position S_1 in Figure 32) when the ecliptic crosses the equator at an angle of $23\frac{1}{2}°$, than at the time of the solstices (S_2), when the Sun's orbit runs parallel to the equator in northernmost or southernmost declinations. In this way a fourfold variation takes place in the annual movement of the mean Sun on the ecliptic. Plotting the position of this mean Sun for midday each day of the year, we obtain a symmetrical lemniscate. It reveals the fundamental form of the analemma.

From the differences which we have described between the mean Suns on the equator and on the ecliptic, and from the resulting curve, it can be seen that the single true solar days are shortest at the time of the equinoxes, and longest at the time of the solstices. The greatest difference, as we have said, totals about one minute. In March and September, therefore, the rhythm of the successive true solar days is somewhat accelerated; in midsummer and midwinter, on the other hand, it slows down.

Finally, taking the step from the second to the third Sun, we reach the astronomical reality. The variable movement of the Sun, due to its eccentric path on the ecliptic (see Chapter 4) causes the form of the true curve of the equation of

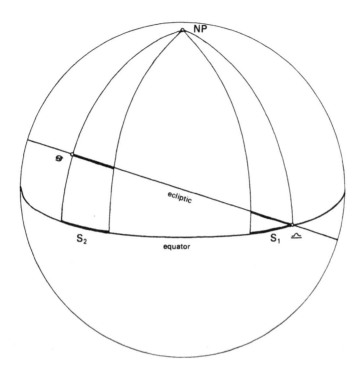

Figure 32. The projection of ecliptical movement on to the celestial equator.

time to be asymmetrical. In accordance with the slower movement of the Sun in the summer months, the monthly midday positions lie closer together in the summer months than in winter, the upper part of the curve becoming narrower, the lower part broader.

The apsidal progression

The Sun's perigee and apogee lie directly opposite one another on the ecliptic, in the Archer and the Twins. In the course of time, however, they do not remain in the same positions, but wander very slowly along the ecliptic in anticlockwise direction: that is, from the Twins to the Crab, and from the Archer to the Goat. In one year the movement totals only 11.6 seconds of arc. A progression of 1° takes about 310 years, and a complete sidereal revolution through all the constellations of the zodiac requires about 110 000 years. The line connecting the apogee and perigee, which is called the *line of apsides*, undergoes a complete turn in this period. This movement is known as the *progression of the line of apsides*.

Through the interplay of the progression of the line of apsides with the precession of the solstices and equinoxes (which we have yet to discuss) apogee and perigee return after a period of 21 000 years to the same position within the

61

seasons, and therefore also to the same position relative to the solstices and equinoxes (Figure 33). The period of 21 000 years is one tropical revolution of the line of apsides.

At the present time the apogee falls in the summer and the perigee in the winter. However, 10 500 years earlier and later we find the situation reversed, so that, for example, the Sun is then nearest the Earth in the summer.

The curve in Figure 31, which could be derived purely from astronomical observations, shows, in addition to the clear asymmetry between the upper and lower parts, a slight asymmetry between right and left. This effect is due to the positions of the Sun's apogee and perigee, which at present lie about 10° from the solstices (Figure 33). This relationship changes gradually. In AD 1250 the line of apsides coincided with the solstices, and will return again to the same position after the course of about 21 000 years.

This period brings about a continual, but very slow change in the form of the curve of the equation of time. A complete cycle through all the changes in symmetry takes place over about 21 000 years. This is illustrated in Figure 34 in eight equal phases. As can be seen, the form of the curve was laterally symmetrical around AD 1250, when the Sun's apogee coincided with the summer solstice, whereas today the form is in the process of becoming more and more tilted. A half-period earlier, that is, around 9000 BC, when the apogee coincided with the winter solstice, the curve was also symmetrical, but with the smaller loop below. In between these two extreme positions this lateral symmetry decreases, and a vertical symmetry of the size of loops takes its place.

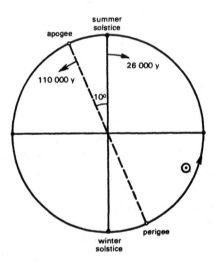

Figure 33. Progression of the line of apsides and of the seasons during the Platonic year.

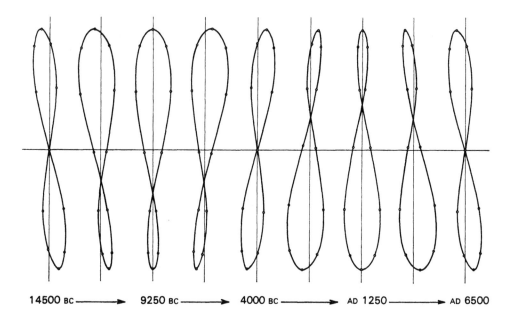

14500 BC ———▶ 9250 BC ———▶ 4000 BC ———▶ AD 1250 ———▶ AD 6500

Figure 34. Metamorphosis of the analemma.

The structure and transformation of the curve of the equation of time arise out of the interplay between the course of the day and the course of the year, together with the long periodic cycles of the Platonic year and the movement of the apogee and perigee.

Sidereal day	Mean solar day		True solar day
23ʰ 56ᵐ 4ˢ	24ʰ		24ʰ ± 28ˢ
Constant	Constant		Variable
sidereal	normal		sundial
clock	clock		clock
	I Sun	II Sun	III Sun
Revolution	Mean Sun on	Mean Sun on	True Sun on
of fixed	equator:	ecliptic:	ecliptic:
stars	constant speed	constant speed	variable speed

Equation of Sun's centre
(maximum value ± 2°)

Equation of time
(maximum value ± 16m, ± 4°)

63

Local time and zone time

Just as sunrise and sunset wander over the Earth, so too midday and midnight. For all points along a line of latitude the Sun culminates at a different time, causing the local times of these points to be different. At a latitude of 50° N or S an east-west difference of 18 km (11 miles) corresponds to about a minute time-difference; 300 m (1000 feet) to about 1 second.

In earlier times clocks were checked against sundials, which naturally showed local time. The railway brought 'rail time' which was valid for a certain route. With increasing speeds and distances, the standard times were introduced in order to establish a uniform time for larger regions. Later the Earth was divided into 24 time zones, each 15° of longitude corresponding to one hour difference. The boundaries are modified by geographical and political factors.

Western European Time is orientated along the 0° meridian running through the Royal Observatory at Greenwich, and is therefore commonly known as *Greenwich Mean Time* (GMT). Central European Time is one hour later, and centred on 15° E. In North America, from Alaska to Newfoundland, there are 8 time zones. *Eastern Standard Time* (EST) is 5 hours behind GMT and is based on 75° W. In addition, some countries have Summer Time or Daylight Saving Time for part of the year, when clocks are set one hour later. This means that civil life begins an hour earlier, making use of the early morning daylight.

Australia has three time zones. The eastern states are 10 hours ahead of GMT, Northern Territory and South Australia are 9½ hours ahead of GMT (giving only half an hour difference between them and their eastern neighbours), and Western Australia is 8 hours ahead of GMT. In addition the southern states have daylight saving time during their summer, thus varying the civil time difference between London and Sydney from 11 hours in January to 9 hours in July.

For astronomical purposes *Universal Time* was introduced, which is based on the Greenwich meridian.

8. Periodic phenomena of the Sun

Sunspots

Given favourable atmospheric conditions, a number of larger and smaller dark spots can often be seen on the Sun's surface by looking through a darkened glass. They can often be seen on the reddish disc of the rising or setting Sun. Chinese annals contain records of the observations of single sunspots two thousand years ago, and throughout the Middle Ages Arabian and Armenian documents, as well as medieval German chronicles occasionally mention their visibility. With the invention of the telescope they were discovered anew for astronomical observation.

8. PERIODIC PHENOMENA OF THE SUN

In the years 1610/11 they were seen almost simultaneously in three different countries: in Italy by Galileo; in Germany by Johann Fabricius (at Emden) and Christoph Scheiner (at Dessau); and in England by Thomas Harriot. In 1842, after observation over a number of decades, the German amateur astronomer Schwabe was able to establish the eleven-year periodicity in the size and frequency of sunspots. The average value of this period is 11.11 years. The individual periods, however, vary in length, and in exceptional cases consecutive spot-minima or spot-maxima can be separated by as few as 7, or as many as 17 years. The rate of increase and decrease is irregular. On the average, the period from minimum to maximum is $4\frac{1}{2}$ years, and period from maximum to minimum $6\frac{1}{2}$ years.

In the years of spot minimum the Sun appears free of spots for many days. In 1954, for example, a year of spot-minimum, 248 days were completely free of spots. In the years following, the number and size of the spots increased with unusual strength, and in the year 1957 reached an hitherto unequalled maximum.

Rudolf Wolf was able to trace sunspot activity as far back as 1610. The years of spot-maximum and minimum can be read directly from the graph (Figure 35), showing the activity from 1700 to 1973.

The individual sunspots wander over the Sun's disc in slightly less than 14 days from the eastern to the western edge of the Sun, where they disappear. They often appear again two weeks later on the eastern edge. This indicates that the Sun rotates around its own axis in roughly 27 days. The life-span of the single sunspots, which undergo constant change in size and form, is extraordinarily variable. Many disintegrate within a few days of their birth and disappear completely soon afterwards. Very large spots are able to exist over a period of several Sun-rotations, and can attain an age of a few months.

At the end of a minimum the first spots of a new cycle appear at a solar latitude of about ± 35°.* In the following years the chief spot regions wander from above and below towards the equator, and at the time of maximum reach a latitude of ± 10°. The last spots of a cycle lie about 5° to the north and south of the solar equator. The movement towards the equator takes place quickly at first, and then much more slowly after the maximum. As the last spots of a period are disappearing near the equator, the first small spots of a new cycle can already be seen in the higher solar latitudes.

The rotation of the Sun

From the movement of the sunspots over the Sun's surface the solar equator and the lines of latitude and longitude are derived, giving in turn the position of the Sun's axis of rotation. The speed of rotation of the spots decreases with increasing

---

*The position of a point on the Sun's surface can be indicated in terms of latitude and longitude, as on the Earth.

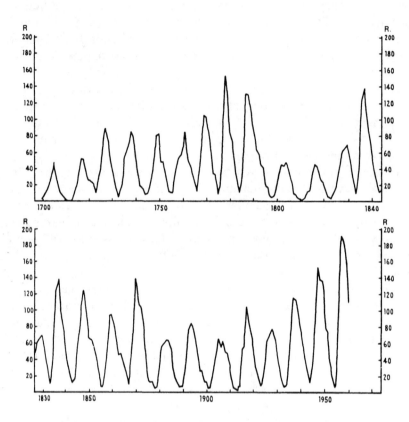

Figure 35. Sunspot frequency 1700–1975. R is the relative number of sunspots.

distance from the solar equator. Other telescopic objects on the Sun's surface also show that the latter does not have a regular, uniform rate of rotation: different parts rotate at different rates. The so-called sidereal rotation, after which a certain point on the Sun has returned to the same position relative to the fixed stars, is roughly 25 days on the equator, but at 35° latitude already 27 days, increasing to about 34 days at 80°. Seen from the Earth, the same sunspot appears again in the same position on the Sun's disc at the end of one synodic Sun-rotation. The latter requires in each case about two days longer than the times given above.

Since the Sun's equator is inclined by $7\frac{1}{2}$° to the plane of the ecliptic, the paths of the sunspots appear differently orientated at different times of the year, owing to the changing relative positions of Sun and Earth (Figure 36). At the beginning of June, and again at the beginning of December, they appear on the Sun's disc as straight lines. The plane of the Sun's equator falls directly in our line of vision, thus passing through the Earth. In the first half of the year the Earth stands below the Sun's equatorial plane, reaching its lowest position on March 6. During this

66

8. PERIODIC PHENOMENA OF THE SUN

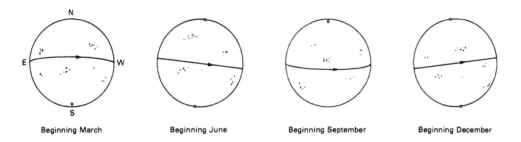

Beginning March   Beginning June   Beginning September   Beginning December

Figure 36. The position of the solar equator and poles (°) in the different seasons. Arrows indicate the direction of movement of the sunspots.

time the spots appear to move along arcs curved upwards, towards the north. Conversely, the paths of the spots are most strongly bent towards the south in the late summer (maximum September 6). At this time of year the Earth is situated above the Sun's equatorial plane.

The different positions of the Sun's equator during the course of the year are illustrated in Figure 37 in eight stages. If we follow the mid-point of the portion

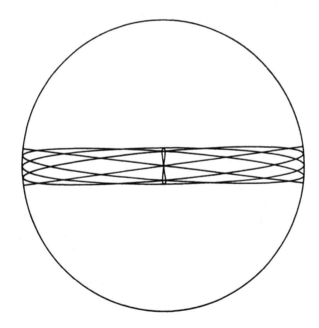

Figure 37. The annual displacement of the solar equator on the Sun's disc and the looping movement of the point nearest the Earth on the solar equator.

Figure 38. Form of corona at sunspot minimum (simplified).

Figure 39. Form of corona at sunspot maximum (simplified).

of the equatorial arc facing the Earth (at the same time the nearest point on the solar equator to the Earth), we find that in the course of a year it describes a symmetrical lemniscate on the Sun's disc. The mid-points of other arcs of solar latitude describe similar annual curves in the form of asymmetrical lemniscates, similar to the curve of the equation of time (see Chapter 7).

The Sun's axis of rotation also assumes different positions relative to the Earth during the course of a year. As can be seen from Figure 36, it describes a double cone, open at both ends by an angle of $2 \times 7\frac{1}{2}°$. The Sun's north and south poles are thus alternately inclined towards the Earth for periods of six months.

Flares, protuberances and the corona

Besides the spots, which appear like pores on the surface of the Sun, other phenomena can be observed with the help of various instruments. In our present study, however, they can only be briefly mentioned.

The entire surface of the Sun appears finely granulated in light and shadow. The individual 'grains', however, appear and disappear in continual flux. Very bright veinlike formations are known as *flares*; they appear in the vicinity of the sunspots and can best be observed on the edge of the Sun. *Protuberances* are flame-like formations, bright red in colour, which appear to be hurled away from the Sun with great force, and then sucked back again with equal energy. The Sun's surface is never at rest; all is in continual motion, in complete contrast to the hardened, unchanging face of the Moon.

During a total eclipse of the Sun a pale white aura of light becomes visible around the darkened disc. This *corona* is most intense on the edge of the Sun and disperses gradually in the surrounding space. Especially in the vicinity of the poles concentrated radial structures can be seen. The corona changes its form and intensity during the course of the sunspot cycle. In Figures 38 and 39 the typical forms at spot-minimum and maximum can be seen; the symmetrical form at sunspot maximum is also considerably brighter than the asymetrical form during reduced solar activity.

9. The precession of the equinoxes and the Platonic year

Just as the Sun alters its position in the zodiac from day to day, a similar displacement occurs, only on a far slower scale, with respect to its annual positions (equinoxes and solstices). The Sun advances roughly 1° along the ecliptic each day; an equal change in its annual position requires 72 years. (In one year this progression comes to about 50″.) A complete revolution through all the zodiacal constellations takes one year in the former case; while in the latter, it requires about 26 000 years (72 × 360 = 25,920), only moving in the opposite direction. This period is called a Platonic year.*

The Platonic year comes to expression in the fact that a given point of the Sun's annual course, as for example the vernal equinox, does not always lie in the same constellation of the zodiac, but moves gradually, over the centuries, into new positions. Figure 40 illustrates these changes in six successive stages for the time from about 7000 BC to AD 3500.

By emphasizing a single cardinal point within the whole movement of the Sun's annual positions, we can speak of the precession of the vernal equinox through the whole zodiac in 26 000 years. In the pre-Christian millennia the vernal equinox moved successively from the Crab to the Twins, the Bull and the Ram; today it lies in the Fishes. Around AD 2500 it will pass into the constellation of the Waterman (Figure 41). More information on the position of the vernal equinox near the brightest and most important stars of the various constellations is given in Table 2.6 in the Appendix.

The passage of the *vernal equinox* from the centre of one constellation to the centre of the next occurs in one twelfth of 25 920 years, that is, in approximately 2160 years. After the completion of one such period, or 'month' of the Platonic year, the autumnal equinox and the summer and winter solstices also lie in a new, characteristic orientation.

*So-called after the Neoplatonic school in Alexandria. During Plato's lifetime the precession of the equinoxes does not appear to have been known.

If, conversely, we regard the vernal equinox as a fixed point, and consider the displacements of the constellations relative to this point over a period of 26 000 years, we find that they show the character of a greatly lengthened yearly cycle. This becomes most evident if we disregard the daily rotation of the fixed stars. Seen from a given place on the Earth, each constellation of the zodiac becomes first a spring constellation, then a summer, autumn and winter constellation successively, thereby passing through the same cycle as the Sun in the course of a normal year. At the same time the length of visibility changes for each constellation in the course of a Platonic year. At the beginning of the Egyptian cultural epoch around 2900 BC, for example, the constellations of the Bull and the Scorpion were at the vernal and autumnal equinoxes, respectively. Both constellations rose and set on the same points of the horizon, directly east and west; similarly, they rose to the same height of culmination in the southern sky. Since that time the Bull has become more and more a winter constellation, whose daily arc carries it to a high culmination, whereas the Scorpion, as a summer constellation, now describes a low arc near the southern horizon.

The remaining constellations above and below the ecliptic also take part in this displacement. The stars of one half of the celestial sphere (centred at the vernal equinox) describe diurnal arcs which, in the northern hemisphere, are growing larger, while those of the other half (centred at the autumnal equinox) describe

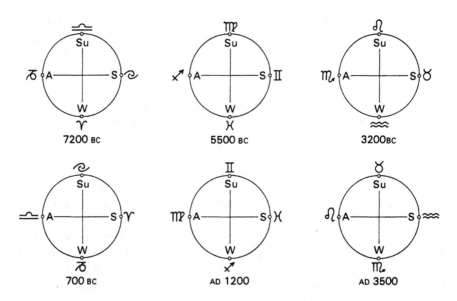

Figure 40. Displacement of the zodiacal constellations with respect to the vernal equinox (S), the summer solstice (Su), the autumnal equinox (A) and the winter solstice (W) in the course of the Platonic year. The years given are those in which the vernal equinox stands in the centre of the respective constellations.

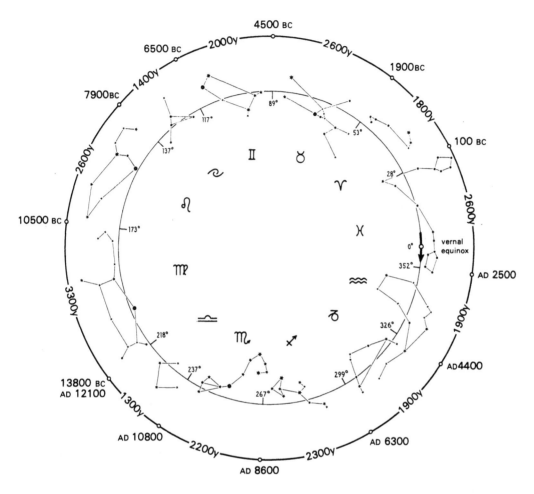

Figure 41. The precession of the vernal equinox in the Platonic year. The degrees on the inner circle and the dates on the outer circle indicate the passage of the equinox from one constellation to the next; the numbers on the outer circle give the approximate length of time which the equinox requires to pass through each respective constellation.

arcs which are becoming smaller. After half a Platonic year, for example, the constellations south of the Archer and the Scorpion, such as the Centaur and the Wolf, which at present never fully rise above the horizon in Europe, will be visible on winter nights in the southern sky. The Archer and the Scorpion, in turn, will by that time have risen to the present position of the Bull and the Twins, our summer constellations.

In the southern hemisphere the Centaur and the Wolf will, during the same period, describe lower and lower arcs over the northern horizon. Conversely, Orion and the Great Dog with Sirius below the Bull and the Twins will become

71

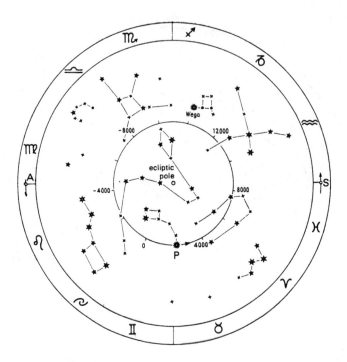

Figure 42. The movement of the northern sky with the celestial north pole (P on the circle) around the ecliptic pole in the course of the Platonic year. The celestial pole moves in the same direction as the vernal (S) and autumnal (A) equinoxes.

invisible in the northern hemisphere and will rise to more and more northern culminations in the southern hemisphere, whereby the Great Dog will even become circumpolar.

Around the celestial north pole the displacement of the constellations during the Platonic year is expressed in the gradual movement of the present Pole Star away from its position near the Pole. Other stars take its place: Cepheus, the Swan and the Lyre (Vega).

The overall movement of the fixed stars during a Platonic year corresponds to a complete rotation of the fixed-star sphere around the ecliptic axis, whose north pole lies in the constellation of the Dragon. Figure 42 illustrates the movement of the celestial north pole (P) and the vernal (S) and autumnal (A) equinoxes relative to the fixed stars and the ecliptic north pole. Conversely, the whole of the fixed-star sphere can be imagined revolving clockwise along the pole-circle.

The displacement of the fixed-star sphere in the Platonic year has been shown to be the reflection of a conical rotation of the Earth's own axis. This *precessional* movement in 26 000 years causes the Earth's axis to point successively towards

9. THE PRECESSION OF THE EQUINOXES

different parts of the heavens, whereby the celestial north pole describes a circle around the ecliptic north pole.

With respect to these phenomena, we have two possibilities for determining the length of a solar year. We can speak either of a *sidereal year*, which is based on the Sun's course through the constellations of the zodiac, or of a *tropical year*, after which the Sun returns to the same position relative to the solstices and equinoxes. Owing to the precession of the equinoxes, the sidereal year is 20 minutes longer than the tropical year.

1 sidereal year = 365.2564 days = 365^d 6^h 9^m

1 tropical year = 365.2422 days = 365^d 5^h 49^m

The traditional terms for the northern and southern tropic circles have been retained by neglecting this slight discrepancy, which only very slowly amounts to a noticeable difference. Today we still speak of the Tropic of Cancer (Crab) at the summer solstice and the Tropic of Capricorn (Goat) at the winter solstice. This in fact corresponded to the reality in Greco-Roman times, about two thousand years ago. At present the Sun reaches its extreme culminations in the constellations of the Twins and the Archer.

The Moon

10. Synodic and sidereal revolution of the Moon

In its cycle of changing forms, the Moon gives to each night a different character. From the most ancient times man devoted great attention to the course of the Moon's phases, in order to determine time or count days. From the celestial script of the Moon he derived his calendar. In the Egyptian myth the god Hermes, or Thoth, who taught men writing, counting and arithmetic, had his dwelling-place on the Moon.

The word 'calendar' (from the Latin *calare*, to call) reminds us that in ancient Rome the first of each month, the *calends*, was proclaimed publicly together with the appearance of the New Moon by a priest from the Capitol. The same usage was current in Asia Minor where the first appearance of the crescent of the New Moon was awaited and then proclaimed to the people with trumpet fanfare.

The *diurnal* movements of the Moon can only be understood in their regular sequence if we regard them as a part of the longer monthly rhythm. In general, the same rule governs the daily movement of the Moon as the fixed stars and the Sun: it rises and sets each day, thereby attaining an upper and lower culmination. The time of day at which these events occur, as well as their location, depend on conditions which change during the course of the Moon's cycle.

The *monthly* phenomena can be divided clearly into two groups: the first describing the Moon's position with respect to the Sun, and the second its passage through the zodiac.

The synodic revolution
The lunar phase period of about 4 weeks, as well as the time, and to some extent the duration of visibility, depend on the Moon's position relative to the Sun. This can be most directly observed in the sequence of phases. The waxing crescent always in the close vicinity of the Sun, can only be seen in the western sky, shortly after sunset. During the waxing phase, the Moon's apparent distance

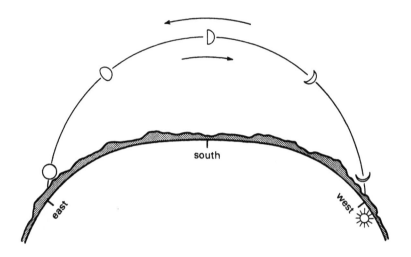

Figure 43. The appearance of the Moon from New Moon to Full Moon just after sunset (shown schematically). The upper arrow indicates the Moon's movement in the zodiac; the lower arrow its diurnal movement.

from the Sun increases from day to day, as it moves eastward. In other words, the difference in ecliptic longitude between the Sun and the Moon increases, namely by about 12° per day.

The waxing Half Moon, or *first quarter*, is reached at an angle of 90°. The Moon stands in a first quadrature to the Sun. In this phase it stands roughly over the south point of the horizon at sunset. About a week later Full Moon follows. The distance from the Sun has increased to 180°: Sun and Moon stand in opposition. At sunset, therefore, the Full Moon is just rising on the eastern horizon (Figure 43).

After a further seven days the distance has increased to 270° or three-quarters of the circumference of the sky. Once again the Moon stands at a right angle to the Sun, now in the second quadrature as the waning Half Moon, or *last quarter*. In this position it rises at about midnight and is visible during the second half of the night. The waning crescent is only visible in the early morning hours, when it appears in the eastern sky before sunrise (Figure 44). Thereupon the Moon passes into the direct vicinity of the Sun and remains invisible for about three or four days. In the middle of this period falls the astronomical New Moon, the conjunction of Sun and Moon.

The New Moon rises and sets almost simultaneously with the Sun; during the day it stands above the horizon. The waxing Half Moon is visible from about midday to midnight, as the regent of the first half of the night. The waning quarter, conversely, dominates the second half of the night and does not set until around noon. The Full Moon is visible during the whole night from evening until morning.

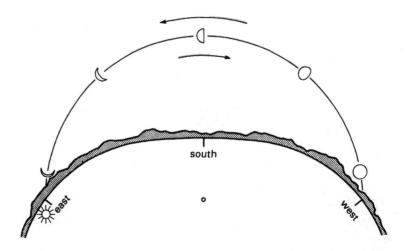

Figure 44. The appearance of the Moon from Full Moon to New Moon just before sunrise (shown schematically). The upper arrow indicates the Moon's movement in the zodiac; the lower arrow its diurnal movement.

Phase	Rising	Culmination	Setting	Visibility
☽ First quarter	midday	sunset	midnight	evening, first half of night
🌝 Full Moon	sunset	midnight	sunrise	whole night
☾ Last quarter	midnight	sunrise	midday	morning, second half of night
🌚 New Moon	sunrise	midday	sunset	invisible (over horizon in daytime)

The Moon grows in brightness and in length of visibility from the evening, from the west, as it were, into the night. As Full Moon it dominates the whole night and at the same time reaches its maximum brightness. The waning Moon decreases towards the east, into the morning.

The Moon has no light of its own; it reflects the sunlight like a mirror. At the time of Full Moon its illuminated half faces towards the Earth; at the time of New Moon it is turned away, as the Moon then stands in the same general direction as the Sun, only nearer to the Earth, so that its far side is illuminated. In the transitional positions we only see a part of the illuminated hemisphere. This appears as a crescent, Half Moon or gibbous* form, which waxes or wanes as the boundary between light and shadow wanders over the Moon's face (this boundary is called the terminator). The fastest increase or decrease of the phases occurs at

*The term 'gibbous' describes the imperfect ellipse of the Moon between the half and full phases.

76

the time of Half Moon; the change is slowest at syzygy, that is, either Full or New Moon. At syzygy, however, the fastest change in brightness occurs. This means that, while there is little difference in phase between the Full Moon and the Moon on the following day, there is a considerable decrease in brightness.

As the illuminated side of the Moon always faces the Sun, it shows us the latter's position below the horizon. We may obtain its direction by following the line perpendicular to the centre of the Moon's light-shadow boundary. This direction always falls within the zodiac, thus giving a new criterion for determining the latter's position.

If we take into consideration that the Moon is not a resting, but a moving mirror of the Sun's light, we may discover yet another fine differentiation in the course of the phases. It is clear that at the time of New Moon, when the angle between Sun and Moon is smallest, the Moon is not only spherically closest to the Sun, but also spatially. It enters its perihelion at the moment when it comes between the Earth and the Sun as New Moon, and conversely, it reaches its aphelion at the time of Full Moon, when it is on the opposite side of the Earth. During its waxing phase it is continually moving away from the Sun, and during its waning phase, towards the Sun. This change in the distance is most rapid at the time of quadrature (Half Moon), and slowest at syzygy (Full or New Moon). We find, therefore, a fine polarity in the Moon's dynamics while waxing and waning.

The pale-blue light which imparts a dim illumination to the whole Moon-disc at the time of the narrow crescent, can best be observed in springtime in the evening, or in autumn in the morning. Leonardo da Vinci was the first to recognize that this illumination of the Moon's shadow-side is due to reflected sunlight from the Earth. It is now called *Earth shine*, its colour being due to the blue of the Earth's oceans.

The sidereal revolution

The Moon's revolution through the zodiac affects its daily arc of movement across the sky. The complete cycle from a higher to a lower culmination and back again, which the Sun completes in a year, requires only 27 days for the Moon. With respect to the southern horizon, therefore, we observe a fortnight's long spiralling ascent of the daily circles of movement of the Moon as it passes from the lower to the higher constellations of the zodiac, from the Archer to the Twins. Then follows a fortnight's spiralling descent, from the Twins to the Archer.

It therefore depends on the Moon's position in the zodiac at what height it will culminate and where it will rise and set. The times of rising and setting are also influenced. The Moon's daily progression from west to east in the zodiac causes a variable delay in the rising and setting times from day to day. In Great Britain the period of delay varies from about 15 minutes (Moon rising in the Fishes) to about 1½ hours (Moon rising in the Virgin). For the setting, the converse is true. Further north the difference is more extreme, while at the equator it remains nearly constant at 52½ minutes. In the southern hemisphere the relationship is reversed: the delay in rising is greatest when the Moon is in the Fishes, and least when it is in the Virgin.

These differences in the delaying of the rising or setting time are dependent upon the position of the zodiac with respect to the eastern or western horizon. Figure 45 illustrates the situation for the Moon setting in the Fishes. Let us suppose that the Moon is standing at the vernal equinox as it is setting; it is 18ʰ local time. A day later the Moon stands 13° farther along in the zodiac, and therefore no longer coincides with the equinox on the horizon. The arrow indicates the movement of the Moon in one hour. The latter will therefore not set until around 19.30. Figure 46 illustrates the corresponding situation in the Virgin. In this case the Moon will set the following day at about 18.20. These relationships are reversed for the moonrise. At the time of the summer and winter solstices, where the ecliptic runs parallel to the celestial equator, the delay in rising and setting times attains its average value of 52½ minutes.

As has been mentioned in Chapter 6, the same geographical laws of rising and setting apply to the Moon as to the Sun (compare Figures 26 to 30). The strongest *'scissoring'* displacement of the time of moonrise takes place at the time of steepest rising, when the Moon is in the Fishes. This occurs in such a way that the time-span between two successive moonrises is shortened. The reverse is true when the Moon stands in the descending part of the ecliptic, most strongly at the autumnal equinox in the Virgin.

The Moon traverses the single constellations of the zodiac in two to four days, depending on their size. Its course through the whole zodiac, known as a *sidereal month* (a month measured with respect to the stars) lasts an average of 27ᵈ 7ʰ 43ᵐ 12ˢ. A lunation, on the other hand (a complete cycle of lunar phases) takes, on the average, 29ᵈ 12ʰ 44ᵐ 3ˢ. This *synodic month* (*synodos*, Greek for 'meeting') gives, for example, the span of time from one conjunction of the Moon with the Sun to the next (from New Moon to New Moon) or from one opposition to the next (Full Moon to Full Moon).

The synodic month is longer than the sidereal, because the Sun progresses by roughly one constellation for every revolution of the Moon. The following New

Figure 45. Moonset in the Fishes.   *Figure 46. Moonset in the Virgin.*

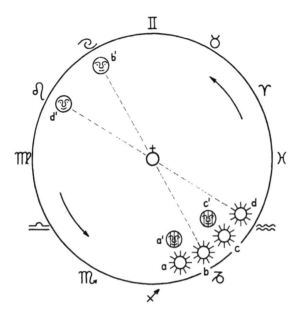

Figure 47. The progression of the New and Full Moons in the zodiac, due to the movement of the Sun. The arrows indicate the direction of movement of Sun and Moon. New Moon (a a'); Full Moon (b b'); New Moon (c c'); Full Moon (d d'), and so on.

Moon, therefore, can only take place after about $1\frac{1}{12}$ sidereal revolutions, or about $2\frac{1}{4}$ days after the completion of one such revolution (Figure 47).

The difference between the synodic and the sidereal months gives rise to a series of phenomena in connection with the seasons. Each successive lunation is a modification of the previous one. Therefore, although the phases always follow the same sequence, no two months are exactly alike.

Each lunar phase has, when it returns, moved along into the next constellation of the zodiac. The New Moon follows directly from month to month the Sun's course through the zodiac. The summer New Moons fall in the upper constellations, the winter New Moons in the lower ones. The Full Moon follows the opposite course through the seasons, as its position is always fixed in the constellation opposite to the Sun. The summer Full Moon stands low and near the horizon in the southern sky. From this time on its daily arc becomes larger, and, after the autumnal equinox, exceeds that of the Sun. The Christmas Full Moon attains the greatest height of culmination, while the Sun describes roughly the arc of the summer Full Moon. As the Sun begins once more to ascend, the daily arcs of the Full Moon decrease, until, after the vernal equinox, they once again become smaller than those of the Sun. The Full Moon in midsummer attains roughly the height of the winter Sun above the horizon, whereas the Sun rises at this time to

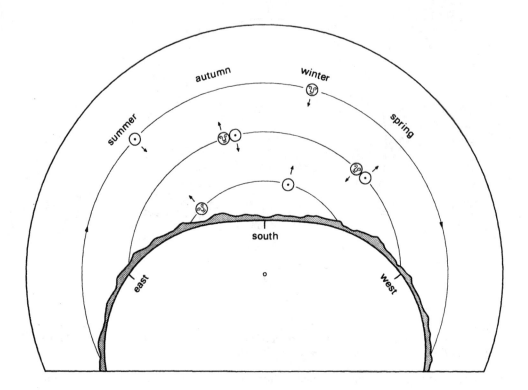

Figure 48. Expansion and contraction of the arcs of Sun and Moon during the course of the year. While the Moon's arcs expand, the Sun's contract.

its highest culmination, similar to that of the winter Full Moon. We therefore find a continual polarity in the rising and sinking of the daily arcs of Sun and Full Moon between summer and winter (Figure 48). The waxing Half Moon attains its highest position in the zodiac in spring, and the waning quarter in autumn. Figure 49 illustrates this interplay of the phases with the higher and lower positions in the zodiac in spring.

Because of the precession of the equinoxes we have had to distinguish between the tropical and the sidereal revolution of the Sun (see Chapter 9). The same applies to the Moon. The latter's course once around the zodiac from an equinox or solstice to the same point again is known as a *tropical month*. It is about seven seconds shorter than a sidereal month.

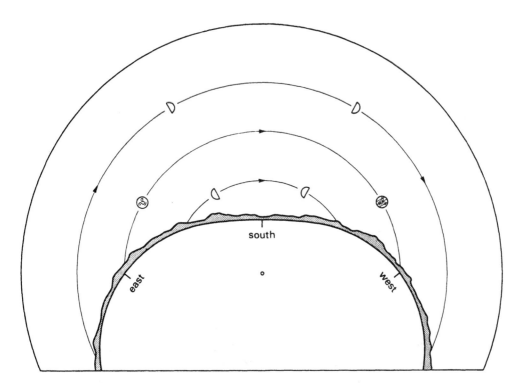

Figure 49. Arcs of movement of the various lunar phases in the spring (simplified).

The Metonic cycle

One solar year comprises roughly 13 sidereal or 12 synodic lunar revolutions. The lunar rhythms, however, are not exactly divisible into a year. Therefore the lunar cycles (both the phases and the positions in the zodiac) are displaced from year to year.

13 tropical months	= 355^d 4^h 20^m
12 synodic months	= 354^d 8^h 49^m

The period of 12 lunations or synodic months is a *lunar year*. It is almost 11 days shorter than the solar year with its 365¼ days. After 33 solar years, 34 lunar years have elapsed, bringing about an equalization of the two rhythms. (The Muslim calendar is based strictly on the lunar year, causing the Muslim New Year to take place 11 days earlier with respect to the seasons each year.)

The Athenian Meton (around 430 BC) calculated the length of time which is required for an equalization of the synodic and the tropical lunar periods. He found that 235 lunations, which take 19 years, are about equal to 254 tropical revolutions:

19 tropical years =	6939^d 14^h 27^m
235 synodic months =	6939^d 16^h 31^m
254 tropical months =	6939^d 16^h 22^m

After the completion of one such *Metonic cycle*, the Moon stands in the same phase on the same date. As far as our calendar is concerned, however, this may vary by one, or even two days, depending on the number of intervening leap years. The Metonic cycle is a rhythm of higher order between the synodic and the tropical month, which results in a whole number of solar years.*

Libration

In its course through the zodiac the Moon always turns the same face towards the Earth, so that the same 'moonscape' (craters, mountains) is always to be seen through the telescope. Only the light and shadow effects are different during the different phases. This shows that relative to the fixed stars the Moon rotates around its own axis once each sidereal month. Relative to the Earth, however, it appears not to rotate. In this respect the Moon displays its character as a satellite, dependent in its motions on the Earth.

A complicated oscillating movement of the Moon, called *libration*, allows us to see slightly more than half the surface of the Moon (59%) at one time or another. The movement has two main components.

Firstly, libration in longitude, which is due to the fact that the Moon does not always move at a constant speed through the zodiac. It is fastest near perigee, when it covers 15° each day, and slowest around apogee, when it covers only about 12°. The maximum resulting libration is about $6\frac{1}{4}$° of lunar longitude.

Secondly, libration in latitude, due to the tilt of the Moon's axis of rotation by about $6\frac{1}{2}$° to its plane of orbit, allowing alternately the north and south poles of the Moon to become visible.

In addition, a further libration in latitude results from the observer's position on the Earth's surface. Observers in different hemispheres see the Moon in slightly different perspectives. This is called diurnal libration, and can amount to as much as 1° of lunar latitude or longitude.

---

*In calculating the date of Easter this 19-year Metonic cycle is used, for at its completion the Easter Full Moon falls again on the same day. The Golden Number gives the placement of each successive year within this cycle.

11. The Moon as a mirror of time

For each time of day and of the year the Moon has its own particular phase and position, making it possible from an observation of the Moon to deduce the momentary cosmic situation. We shall illustrate this with two examples.

Let us suppose that the Moon is standing as a small, strongly inclined crescent just over the east point of the horizon (Figure 50). We can then ask: what is the time of day and the season; what is the sidereal time, and can Sirius be seen?

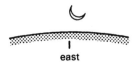

Figure 50. Crescent Moon over the eastern horizon.

The delicate crescent in the morning sky is not far from the Sun. It must be shortly before sunrise. If both luminaries are rising exactly in the east, they must be either in the constellation of the Fishes or the Virgin. But how can we decide which?

The attentive observer will not fail to notice that each of the year's twelve Moon crescents in the morning sky not only stands over a different part of the horizon, but also forms a different angle to the horizon. The placement of the crescent reflects roughly the inclination of the ecliptic, whereby the extreme positions occur in spring and autumn (Figure 51). In Figure 50, therefore, we have the autumn situation in the Virgin. The twelve crescents in the evening sky show a similar differentation. Here, however, the situation is exactly reversed with respect to the Fishes and the Virgin, as a moment's reflection will clearly show.

As the Virgin rises in the east, the Fishes with the vernal equinox are setting

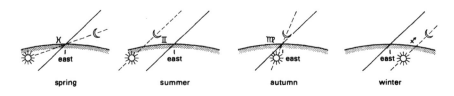

Figure 51. The relative position of the crescent Moon and the ecliptic (– – – –) to the celestial equator (————) over the eastern horizon at different times of year.

in the west. When the vernal equinox culminates in the south it is 0^h sidereal time; 6 hours later it sets below the western horizon. It is therefore 6^h sidereal time.

At this hour the steep upper half of the zodiac arches above the horizon. The constellations of the Bull and the Twins stand high in the southern sky. Below them lie the constellations of Orion and the Lesser and the Greater Dog. Sirius, the brightest star of the latter constellation, is therefore visible in the southern sky. Similarly, the positions of all the other constellations can be found.

In our second example the Moon is rising in its last quarter in the south-east (Figure 52). We shall try to answer the questions: what is the time of day and the season, in which constellation is the Moon, which constellation is culminating, is Regulus above the horizon, and what is the sidereal time?

Only the constellations of the Scorpion and the Archer have their rising-points in the south-east. The Sun is located 90° further along in the zodiac than the Moon in its Last Quarter: in this case, in the Fishes. When the Archer rises in the south-east, the Fishes with the Sun are about one hour past their lower culmination below the north point of the horizon. Therefore, it is just past midnight at the beginning of spring. The vernal equinox is also past its lower culmination: it is therefore about 13^h sidereal time.

When the Archer is rising, the Twins are setting in the north-west. The Virgin is then past its upper culmination. To the west of her the Lion with Regulus is inclined towards the western horizon (compare Figure 53).

Further we can ask, where will the next first quarter, Full and New Moon rise?

The corresponding first quarter is 90° further along in the zodiac than the Sun, and therefore rises with the Twins in the north-east. The Full Moon is opposite the Sun in the Virgin; it will rise in the east. Similarly, the New Moon rises with the Sun in the Fishes. This is the situation in spring.

In the autumn the Moon in syzygy (Full or New) also rises in the east, but the rising points of first and last quarters are reversed. In winter the two quarters

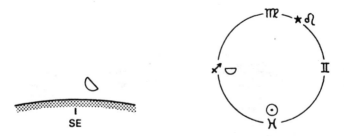

Figure 52. Half Moon in the south-east.   Figure 53. Sun, Moon and Regulus.

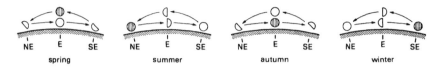

Figure 54. The rising positions of Full Moon, New Moon, first and last quarters in the different seasons.

rise in the east, while the Full Moon rises in the north-east (opposite the Sun in the zodiac), and the New Moon together with the Sun in the south-east. In summer the quarters are similar, but the Full and New Moon are again reversed (Figure 54).

These considerations throw further light on the Egyptian myth mentioned earlier, according to which Thoth, as regent over the ordering of time, has his seat on the Moon.

In different parts of the world, the above considerations for a northern latitude of around 50° must be modified.

12. The nodes and apsides

The nodes

Were the Moon to follow exactly the same course through the zodiac as the Sun, it would pass directly in front of the latter, blocking out its light at each New Moon. This, however, is not the case. As more precise observation shows, the Moon stands sometimes north and sometimes south of the ecliptic, its orbit inclined at 5° to that of the Sun (Figure 55). On the background of the fixed stars, therefore, the orbits of Sun and Moon cross at two diametrically opposite points in the zodiac. These are known as the *lunar nodes*. During the course of each sidereal revolution the Moon crosses the ecliptic once from south to north at the northern or ascending node (☊) and some fourteen days later from north to south at the southern or descending node (☋). Between the nodes it travels alternately for about 14 days in northern or southern ecliptic latitude, whereby its maximum distance from the ecliptic is 5°, or roughly 10 Moon diameters.

The lunar nodes, however, are not fixed, but progress continually along the ecliptic from east to west or clockwise; in other words, in the opposite sense to the planetary motions. This movement totals about $1\frac{1}{2}$° in a month. The Moon's course from one node to the same node again must therefore be shorter than the sidereal month: the *nodal month* (or archaically draconitic month) lasts 27^d 5^h 5^m 36^s. In a year the lunar nodes regress by about 19°, or roughly two-thirds of a

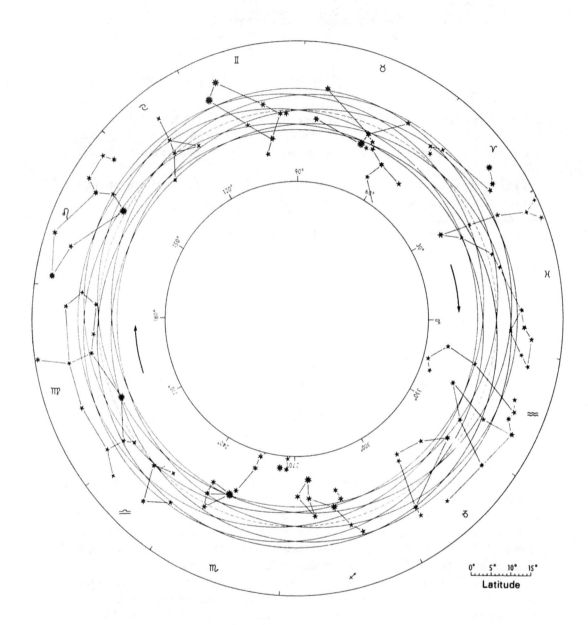

Figure 55. The Moon's orbit in the zodiac and its displacement due to the retrograde movement of the nodes. The displacement, which is of course a continuous movement, takes place in the direction of the arrows.

12. THE NODES AND APSIDES

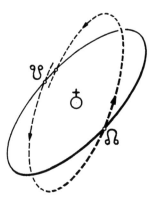

Figure 56. Ecliptic (——) and Moon orbit (– – – –). The arrows indicate the direction of the Moon's movement along its orbit.

constellation. A complete revolution through the zodiac requires 18.6 years (18 years, 7 months). The other points along the Moon's orbit, as for example the points of greatest distance from the ecliptic, also participate in this motion. This is illustrated in Figure 55. The individual positions gradually pass over one into the next. The Moon's path through the zodiac is not a closed line returning again to its starting-point, but rather an open curve, winding like a ball of wool, so that one position lies beside the other (Figure 56).

The four most characteristic situations of the Moon's orbit with respect to the ecliptic may be observed when the nodes coincide with the equinoxes or the solstices (Figure 57). In some years the Moon's orbit rises 5° north of the northern-most point of the ecliptic, and descends similarly below the southernmost point (Figure 57a). In such years, as for example 1968–69, 1987 and 2005–6 the summer Full Moon culminates especially low, and the winter Full Moon especially high. Nine years later the extremes are balanced out. At these times the winter Full Moon stands 10° lower, that is 5° south of the ecliptic at its culmination, and the

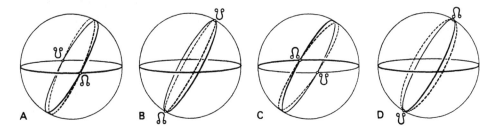

Figure 57. The position of the Moon's orbit (– – – –) relative to the ecliptic (———) when the nodes coincide with the equinoxes and solstices during a nodal revolution of 18.6 years (compare Figure 58). Ascending node (☊); descending node (☋).

summer Full Moon correspondingly higher, 5° north of the ecliptic (Figure 57c). This is the case in the years 1978, 1996 and 2015.

The belt of the sky which the Sun covers during its annual course retains the same breadth year in, year out: it comprises a zone of twice $23\frac{1}{2}$°, or 47°. That of the Moon, however, is constantly changing. If the northern node coincides with the vernal equinox (Figure 57a) the Moon's field of movement overlaps that of the Sun by 5° above and below, measuring 57°. If the northern node is at the autumnal equinox, the Moon's field lies within that of the Sun, and has a breadth of only 37° (Figure 57c). The overall relationship is illustrated in Figure 58. This breathing expansion and contraction of the Moon's field takes place over this period of 18 years and 7 months.

The effects of this rhythm may become clearer if we look at a particular constellation through which the Moon passes in its monthly course, for example the Bull. In 1987 the Moon passes just over the Pleiades, and on towards El Nath, the bright star on the tip of the northern horn. In 1996, however, it passes just

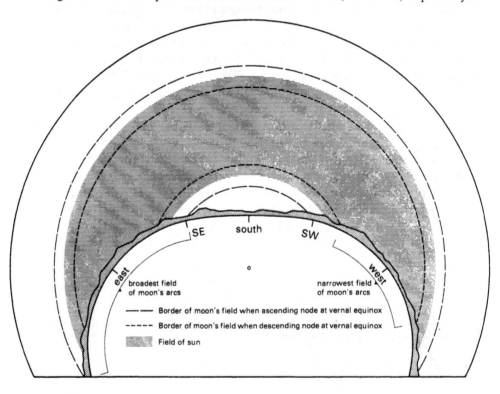

Figure 58. The broadest and narrowest fields of the Moon's arcs, with the position of the nodes corresponding to that of Figures 57a and 57c respectively. The field of the Sun's arcs remains constant.

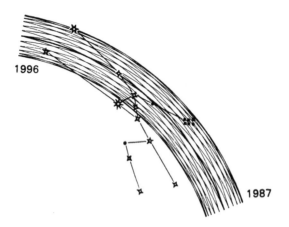

Figure 59. The changing situation of the Moon's orbit in the Bull over a period of 18.6 years. Ecliptic: ————.

over Aldebaran, 10° further south (Figure 59). In its slow rising and sinking movement the Moon's orbit passes through various slanted placements, so that again a kind of 'scissoring' upwards and downwards motion results.

A survey of all possible positions of the Moon's orbit at a particular part of the ecliptic also shows how different the relative placement of Sun and Moon can be at the time of a conjunction. The Moon can, in its extreme positions, pass 5° above or below the Sun, that is, with a separation of about ten Sun or Moon-diameters. If, however, one of the lunar nodes stands at the point of conjunction, the New Moon will pass directly before the Sun, resulting in a solar eclipse.

The oscillation of the Moon's orbit can, with very exact observation, be seen in yet another part of the sky. The celestial poles, in their slow progression through the Platonic Year, do not follow, as we would expect, a smooth curve, but rather a wavy line. This is due to a kind of nodding movement of the celestial poles over a period of 18.6 years (6798 days). The waves have a half-amplitude of 9 seconds of arc, and are directed towards the ecliptic pole. Measured from the latter, the waves are about 6 minutes or arc in length, or about a fifth of the Moon's diameter. The period of 18.6 years is therefore also known after this movement as the *period of nutation*. As seen from the Earth, the whole sphere of the fixed stars appears to participate in this nodding movement.

The apsides

The attentive observer will not fail to notice yet another rhythm in the Moon's movements. Like the Sun, the Moon travels at a varying speed along its orbit. Its daily movement fluctuates between $11\frac{3}{4}°$ and $15\frac{1}{4}°$. At the same time, the apparent

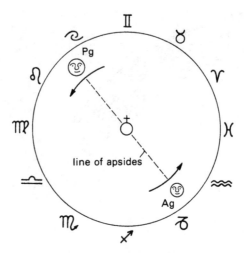

Figure 60. The progression of the Moon's line of apsides through the zodiac with perigee (Pg) and apogee (Ag).

size of the Moon's disc, which is on the average 31′ 06″, varies between 29′ 23″ at minimum and 33′ 31″ at maximum.* The ratio between its smallest and greatest size is about 7:8. This difference becomes particularly noticeable when the Full Moon takes place at one of the extremes. The total apparent area of the disc of the Full Moon at minimum is only about three-quarters of the area at maximum. Both of these variations, that of the fluctuating speed and that of the changing size, are due to the Moon's being sometimes nearer the Earth and sometimes farther away.

The Moon's closest point to the Earth is called *perigee* and its farthest point *apogee*. These positions, also known as the *apsides* (plural of apsis) constantly change their situation in the zodiac. The *line of apsides* (the line joining perigee and apogee) moves forward (anticlockwise) through the zodiac by over 40° each year, or by an average of about 3° per lunar revolution (Figure 60). The Moon's progression from one perigee to the next will therefore be longer than a sidereal month: the *apsidal month* lasts 27ᵈ 13ʰ 18ᵐ 33ˢ. The displacement in a year, however, is not the sum of equal monthly progressions, but is the result of a complicated back and forth oscillation of the apsides, whereby the forward motion prevails.

The oscillations do not take place simultaneously, but alternately, so that the apsides are not always directly opposite one another in the zodiac, but are continually falling behind and overtaking these positions. The retrograde motion of the

*These figures arae true at the poles of the Earth. The respective figures for the equator are: 31′ 37″, 29′ 51″ and 34′ 08″. The ratio is even closer to 7:8.

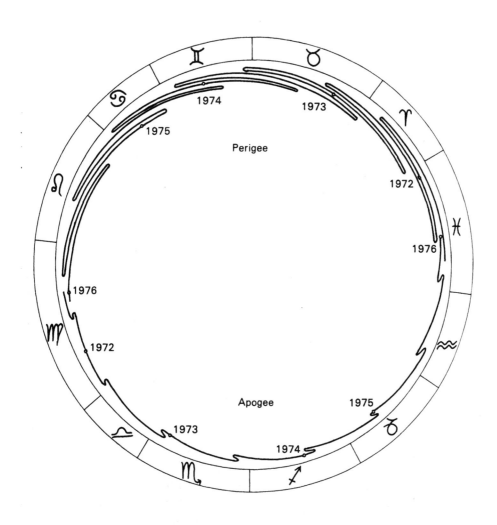

Figure 61. The movement of the lunar apsides in the years 1972–76.

perigee (roughly 40°) is very much larger than that of the apogee (roughly 2° to 3°), meaning that the former moves much more quickly than the latter against the fixed star background. This polarity comes most strongly to expression in the vicinity of the retrograde motion, at which time the perigee can regress by more than 30° in a single month, whereas the apogee moves for up to four months

91

within a field of only about 3°. Figure 61 illustrates this movement for the years 1972 to 1976. The representation is strongly schematic, showing the retrograde motion as a loop-form, in order to make it visible in the drawing. In reality, they have no semblance to the planetary loops, as they always remain within the plane of the lunar orbit. Moreover, it must be borne in mind that the curves in the drawing are obtained by connecting all points at which the Moon actually stands in perigee or apogee. In between these positions the apsides cannot in the same sense be thought of as *real* points; they elude our investigation, as it were, and only 'materialize' again when the Moon has reached them in its orbit.

A complete revolution of the apsides through the zodiac takes place in roughly 9 years (more precisely, 8.85 years, or $8^y 310^d$). Figure 62 illustrates the resulting situations of the Moon's orbit for a longer period of time.

If the Moon's exact position against the fixed stars is measured simultaneously from two widely separated stations on the Earth, the two measured positions show a certain discrepancy, due to the difference in the direction of observation. (See also the diurnal libration, described on page 82.) Such measurements can serve as the basis for a calculation of the Moon's distance from the Earth. Its mean distance is found to be about 30 Earth diameters, or 384 400 km (238 860 miles); the maximum distance is 406 740 km (252 740 miles), the minimum 356 410 km (221 460 miles).

The Moon's true diameter is 3 476 km (2 160 miles), a distance roughly corresponding to the length of the Mediterranean from Gibralter to the Levant, or the distance from Newfoundland to Florida.

By way of summary, the following table may be of help in remembering that certain lunar rhythms proceed in opposite directions in the zodiac:

daily movement of the Moon	east to west
synodic month (phases)	west to east (direct)
sidereal month	west to east (direct)
nodal month	east to west (retrograde)
apsidal month (general)	east to west (direct)
(in detail)	oscillating, alternately direct and retrograde

In addition to these more important lunar rhythms, many finer fluctuations and variations take place which make the exact calculation of the Moon's position and orbit extremely difficult. Every conjunction or opposition of the Sun with a planet has a fine accelerating and retarding effect on the Moon's movement, and changes slightly the position of its orbit. Conversely, it is theoretically possible, from extremely exact measurements of a small segment of the Moon's path, to calculate the positions of all the planets.

In this connection we should note that all the indications for the lengths of the various Moon rhythms are mean values. An individual synodic month, for

92

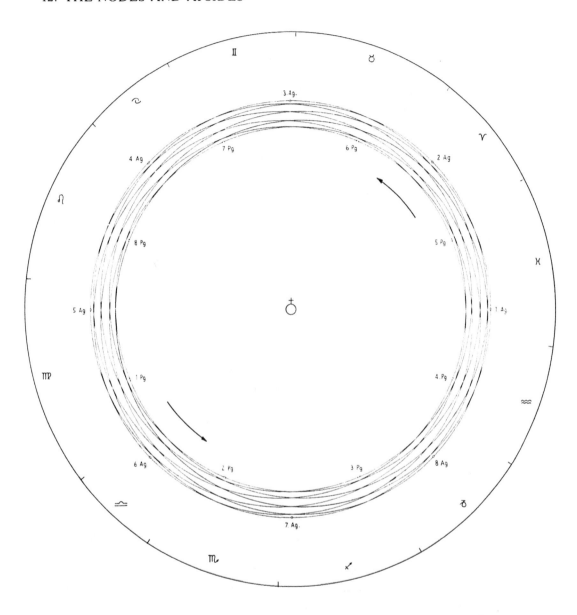

Figure 62. The geocentric orbit of the Moon. The displacement results from the progression of apogee (Ag) and perigee (Pg) in the direction of the arrows. The orbit therefore moves in the sequence: 1.Ag, 1.Pg, 2.Ag, 2.Pg, 3.Ag, and so on. The movement of the apogee from 1.Ag to 2.Ag takes about 2.2 years. The intermediate positions must be imagined, as the real movement is continuous.

example, can deviate up to seven hours from the mean. Its mean length, nonetheless, was already calculated to within a few minutes by Hipparchus and Ptolemy.

Altogether, several hundred such minute periodic fluctuations in the Moon's movement may be distinguished. In the language of celestial mechanics they are called 'perturbations'. Since the earliest times they have given astronomers the greatest difficulties in comprehending the Moon's orbit. Theories pertaining to the Moon had constantly to be extended and corrected, particularly on account of the increasing exactness in the methods of observation. The French astronomer Charles Eugène Delaunay has left us a comprehensive description of the Moon's orbit. In the first volume of his 1860 work *Théorie du mouvement de la lune* he made public, alongside other gigantic formulae, his so-called 'perturbation function of the Moon.' It is 138 quarto pages in length. The second volume (1867) contains the formulae for the calculation of the Moon's longitude (173 pages), latitude (155 pages) and of the reciprocal value of the radius-vector Earth-Moon (16 pages). Altogether 482 pages. But even so, this study by no means exhausts the Moon's movements, as new perturbances are constantly being discovered. Delaunay's calculations are no longer in general use. Modern astronomical ephemerides are calculated using modifications of the lunar theories of the American astronomer G. W. Hill (1838–1914).

The Moon, whose surface appears as a dead and hardened cinder, possesses the most complex and varied rhythms of movement of all the luminaries. It eludes man's attempt to grasp its movements by means of rational, numerical calculations.

13. Eclipses of the Sun and the Moon

Solar eclipses
Eclipses of the Sun and the Moon are the result of the relative movements and positions of Earth, Moon and Sun. A solar eclipse takes place when a New Moon occurs in the immediate vicinity of one of the lunar nodes. Only near the crossing-points of the orbits of Sun and Moon can the latter, in passing, partly or entirely cover the Sun's disc, thereby eclipsing its light. If the meeting occurs very close to the node, the eclipse will be central: either total or annular.* If the New Moon is more than $18°4$ from the node an eclipse cannot occur; if it is less than $15°4$ from the node an eclipse must occur. These two values are called (respectively) superior and inferior ecliptic limits. If the separation is between these values an

*Coming from the Latin *annulus*, meaning 'little ring'.

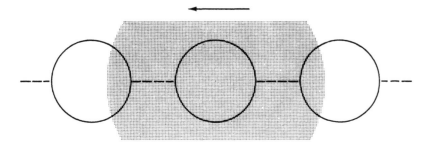

Figure 63. The passage of the Moon through the central shadow of the Earth during a lunar eclipse. The arrow indicates the direction of the Moon's movement.

eclipse may occur, depending on the exact relative positions of Sun, Moon and Earth.

The apparent size of the Sun's and the Moon's discs is almost exactly the same ($\frac{1}{2}°$); it fluctuates, however, with their changing distances from the Earth. If the Moon is near its perigee, a total solar eclipse is possible; if it is near its apogee, an annular eclipse can occur. In the latter case, the Moon's disc is slightly smaller than the Sun's, leaving a narrow ring exposed around the edge of the Sun.

Lunar eclipses

Eclipses of the Moon, which in many respects display a polarity to the solar eclipses, occur at Full Moon, when Sun and Moon stand in opposition. If the Full Moon occurs near a node (the respective superior and inferior ecliptic limits are 12°1 and 9°5) the sunlight which normally illuminates the Moon will be partially or totally obliterated by the Earth; the Moon enters the Earth's shadow. Should it plunge completely into the shadow, a total eclipse will result; if it only grazes it, the eclipse will be partial. The Earth's shadow has, at the distance of the Moon, a diameter of 83 ± 8 minutes of arc, which is nearly three times the diameter of the Moon. The variation is caused by the changing distance between the Earth and the Moon, which has been discussed in the previous chapter.

Each total lunar eclipse begins with the entrance of the eastern edge of the Moon into the shadow (first contact). The eclipse, which is always partial in the beginning, 'waxes' for roughly one hour before it enters totality (second contact), which can, in turn, last as long as 1^h 40^m. The eclipse becomes partial again when the Moon's eastern edge emerges once more into the light (third contact), and ends when its western edge leaves the zone of darkness (fourth contact). The total duration of a lunar eclipse can reach 4 hours. The succession of phenomena thereby appears as a kind of distorted recapitulation of the lunar phases, only in inverted sequence (Figure 63). The Moon's darkened disc does not become

95

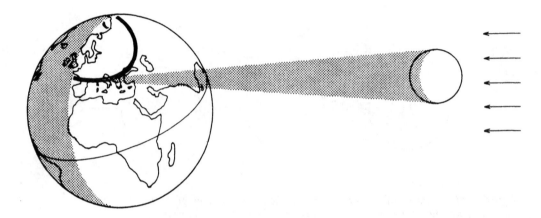

Figure 64. The tip of the Moon's shadow-cone on the Earth and its path during the solar eclipse of February 15, 1961.

completely invisible during totality. Towards the centre of the Earth's shadow it displays reddish-brown tones, and in its outer portions dusky bluish colours.

Contrasts between solar and lunar eclipses

The successive phases of a lunar eclipse can be observed at the same time and in the same form over all those portions of the Earth where the Moon is above the horizon, that is, over an entire hemisphere.

The visibility of solar eclipses, on the other hand, is restricted to more limited areas of the Earth's surface. The totality is only simultaneously visible within a region of up to about 260 km (160 miles). This is the region where the narrow, tapering cone of the Moon's shadow touches the Earth's surface (Figure 64). North and south of this belt is a zone extending several thousand kilometers in each direction where the Sun appears partially eclipsed to various degrees. North of the central region the northern edge of the Sun remains exposed; to the south the southern edge. The whole shadow moves from west to east over the continents

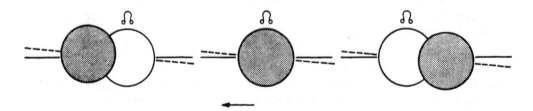

Figure 65. The passage of the Moon in front of the Sun during a total solar eclipse. The arrow indicates the Moon's movement. Ecliptic: – – – –; Moon's orbit: ———

and oceans, tracing a path whose length can reach as much as a third of the Earth's circumference. For any zone of observation over which the totality zone passes, the totality itself lasts only a few minutes. Three to four are already quite favourable; in unusual cases it can last as much as eight minutes.

The passage of the shadow-cone from west to east is dependent above all on the sidereal movement of the Moon, which during the conjunction passes in front of the Sun from right (west) to left (east) (Figure 65). In addition, the diurnal rotation of the Earth moves in the same direction as the shadow, causing the latter to pass more slowly over the Earth's surface than would otherwise be the case. Its minimum velocity (near the equator) is about 35 km per minute (20 miles per minute).

At the beginning of a solar eclipse the Moon touches the western edge of the Sun's disc (first contact), until it either completely covers the latter, or, in the case of an annular eclipse, is completely in front of the Sun (second contact). After a few minutes or seconds of totality the Moon begins to leave the Sun on the latter's eastern edge (third contact), making the eclipse partial until the Moon completely leaves the Sun (fourth contact). The forms of the successive solar eclipse phases thus display the inverted sequence to those of the lunar eclipses. The period of visibility of a solar eclipse from beginning to end at a particular point on the Earth can reach up to 3 or 4 hours. The period of totality over the Earth's whole surface is normally from 2 to 4 hours, although in exceptional cases it can be longer.

In its complete course a total solar eclipse recapitulates in brief the phenomena of the course of a day, but in polar form. As the Moon's shadow laterally approaches the Earth, the first contact always takes place on a point on the Earth where the Sun is just rising. As the shadow proceeds over the Earth's surface, it enters regions where other times of day prevail. The culmination of the eclipse takes place over a region where it is just noon, and the end occurs in a locality where the Sun is just setting. The complete course of the eclipse thus takes the form of a spatially extended 'eclipse-day' which moves from west to east over the Earth's surface.

Annual eclipse periods

Every year there are two *eclipse periods* which occur at intervals of six months, when the Sun crosses one of the two lunar nodes at the ecliptic. Each time a solar eclipse must occur, so that two such eclipses take place each year: one at the northern node, the other at the southern node (see Tables 6.1 and 6.2 in the Appendix). In some years two eclipses take place during one eclipse-period, at two successive New Moons, as for example in 1982 and 2000. As can be seen from Figure 66, both solar eclipses are partial and of brief duration, for the New Moons occur at a considerable distance to either side of the node. In extremely rare instances one of the eclipses can be central. This last occurred in 1787 (June 15

97

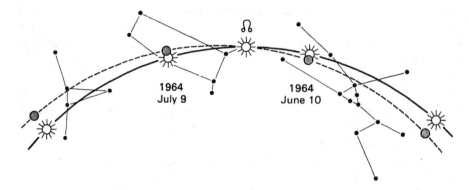

Figure 66. The configurations of two partial solar eclipses, separated by a four week interval. The diameters of Sun and Moon are enlarged.

total, July 14 partial) and will not occur again until 2195. Occasionally, two consecutive periods each produce two eclipses. This occurred in 1916/17 and in 1935. It does not occur again in this or in the next century. Usually every nine or ten years there is an eclipse at the beginning of January and an eclipse at the same node towards the end of December. This is the case in 1982, 1992 and 2000. It results in three or four eclipses taking place in the same year.

Lunar eclipses take place during the same eclipse periods. In accordance with the ecliptic limits discussed at the beginning of this chapter, however, they are less frequent. The reason for this becomes clear through a simple geometrical consideration. In order for a solar eclipse to occur, the Moon, or a part of it, must come between the Earth and the Sun, so that its shadow falls somewhere on the Earth's surface. The region in which this is possible is almost $1\frac{1}{2}$ times the size of the Earth at the distance of the Moon. Conversely, a lunar eclipse requires the Moon or a part of it to enter the cone of the Earth's shadow. At the distance of the Moon this region is nearly equal in size to the Earth. It follows that about $1\frac{1}{2}$ times as many solar eclipses occur as lunar eclipses. A glance at the table of lunar eclipses shows that there are years with two, one and no eclipse. In the years 1980, 1984, 1998 and 2002 no lunar eclipse occurs. In 1982 there were three, the eclipse periods being January, June and December.

As the lunar nodes move round the zodiac over an 18.6 year period, the eclipse periods also move through the seasons, occurring about 3 weeks earlier each year. However, as eclipses can only occur at Full or New Moon, they take place earlier from year to year by complex intervals resulting from the interplay between the nodal and phase rhythms. In Chapter 10 we have seen that the syzygies (Full and New Moon) occur about 11 days earlier each year. They therefore move retrograde (clockwise) through the zodiac. The lunar nodes also move retrograde, but at about double the speed, so that they slowly overtake and pass the syzygies. As a result, the eclipses will take place from year to year about 10 or 11 days earlier over a period of 3 to 4 years before leaping to the previous syzygy, in which case they will occur about 40 (11 + 29) days earlier. In just what intervals this alternation occurs depends on the individual distribution of the syzygies around the node in each case. Examples may be derived from the eclipse tables in the Appendix (6.1).

About every 9 or 10 years the solar eclipses fall in midsummer or midwinter, and 4 or 5 years later at the times of equinox (Figure 67). The eclipses which follow each other in this way from year to year have no further connection with each other. After 18 years the cycle is completed, and each eclipse recurs in cyclical sequence.

The Saros period

The actual length of the eclipse cycle is 18 years, 10 days and 8 hours. It appears to have been known in relatively early times to Chinese and Chaldean astronomers, and is designated as the *Saros period*. At its completion Sun and Moon meet once again at the same lunar node, in very nearly the same configuration.

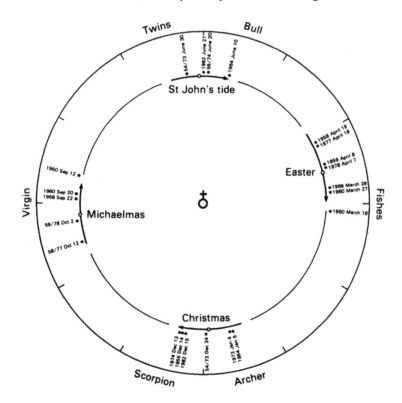

Figure 67. Solar eclipses at the four cardinal festivals from 1974 to 2007.

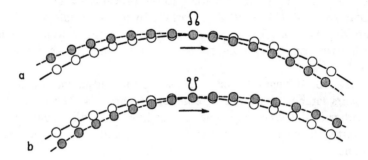

Figure 68. The progression of the eclipses through the region of a node.

The Saros is very nearly a common multiple of the following solar and lunar rhythms, which all play a role in the cycles of eclipses.

The Sun passes each node once in a year. The exact interval between two such meetings, known as the *eclipse year (nodal solar revolution)*, is 346.620 days, which is nearly three weeks shorter than an ordinary year, owing to the annual regression of the nodes. A Saros contains 19 such intervals. In the same period 223 synodic months pass (the true length of the Saros), 242 nodal months and 239 apsidal months, showing that the Moon's distance from the Earth is also the same. We have:

19 eclipse years	= 6585.781 days
242 nodal months	= 6585.357 days
239 apsidal months	= 6585.537 days
223 synodic months (1 Saros)	= 6585.321 days

We can see that after 223 lunations, or a Saros of 18 years and 10 days, a New Moon recurs which induces an eclipse very similar to that which took place 18 years earlier. The slight time difference between the Saros and the 19 eclipse years (a difference of 0.46 days) brings about a very slight change in the new eclipse. The new conjunction of Sun and Moon is therefore slightly displaced with respect to the node. This displacement of $0°.45$ proceeds from east to west. It brings about a continual metamorphosis and development of each related series of eclipses.

Solar eclipse series

Figure 68 illustrates the situation for a solar eclipse at the northern node (a) and for another at the southern node (b). Following this pattern, the eclipses gradually advance through the whole region of the node in steps of a little less than one Sun-diameter for each eighteen-year interval. They develop from small partial eclipses in the beginning (Figure 68, left-hand side) which gradually 'wax' until they become central. After several centuries these again pass over to waning partial eclipses and in the end expire.

100

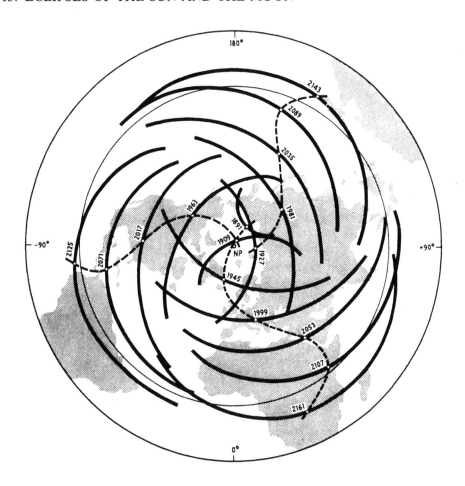

Figure 69. The displacement of the totality zone (———) of an eclipse series from the North Pole southwards. The eclipses of July 9, 1945, and August 11, 1999 belong to this series. The centres of the totality lines have been joined to form 'guiding lines', also called exelegismos curves (– – – –).

The whole course of such a series of organically related eclipses appears as a kind of gigantic summary of a single eclipse in its various stages, with a beginning, a waxing to greatest intensity, waning and finally disappearance.

The whole development of a Saros series of solar eclipses lasts from twelve to fourteen hundred years, and comprises on the average about seventy single members ($70 \times 18 = 1260$). The Saros series of lunar eclipses contain roughly forty-five individual eclipses, accounting for their shorter duration of about eight to nine hundred years ($45 \times 18 = 810$).

Within any eighteen-year period, in turn, an average of forty-two solar eclipses

101

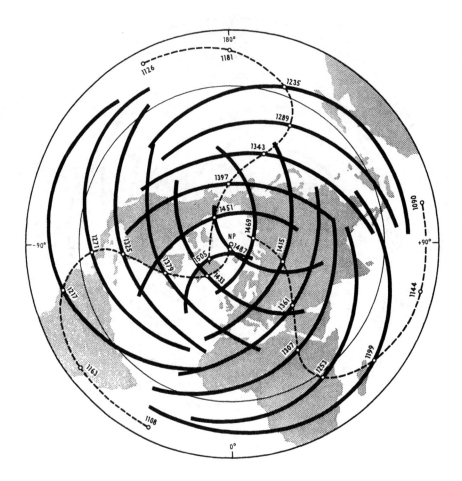

Figure 70. The displacement of the totality zone (———) of an eclipse series moving northwards. Exelegismos curves (– – – –).

and twenty-seven lunar eclipses takes place.* At a given time, therefore, there are on the average forty-two Saros series of solar eclipses and twenty-seven series of lunar eclipses. They all proceed independently of one another, but overlap each other in time, and interweave in rhythmic alternation through the centuries. Regarding a single eclipse we can ask: at what stage of its own Saros series is it? Does it have the greater portion of its life-history before it or behind it? The short

*The number of solar eclipses can vary between 38 and 46; the lunar eclipses between 25 and 29. This variation depends on the whole on a larger rhythm of about 600 years within which the individual Saros periods in turn vary, according to their composition.

partial eclipses are always either young and in their first beginnings, or old and passing away. Extended total eclipses are in the middle of their series. Details can be found in Meeus, Grosjean and Vanderleen's *Canon of Solar Eclipses*.

In the whole of the twentieth century only two new solar Saros series come into being: they appeared for the first time on July 19, 1917 and on June 17, 1928 as small partial eclipses. The next series will begin on July 1, 2011. Five Saros series, however, disappeared: April 8, 1902; September 12, 1931; January 5, 1935; August 12, 1942; and July 22, 1971. In the next century only two cycles end: in 2054 and in 2083.

In the earliest members of a Saros series which develops at the northern node, only the northern edge of the Sun is eclipsed (Figure 68a, left). Such eclipses are only visible near the North Pole. The last members of the same series, in which only the southern edge of the Sun is grazed, can only be observed in the Antarctic. Conversely, all Saros series of the southern node appear for the first time near the South Pole and wander gradually northwards over the Earth's surface, until they disappear in the Arctic.

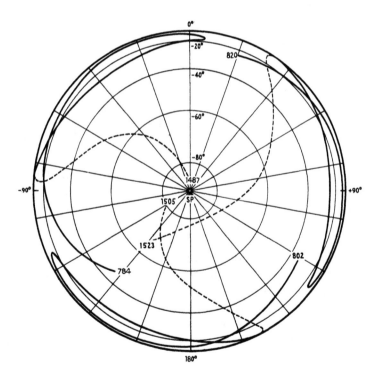

Figure 71. The same exelegismos curves as in Figure 70, but seen from the South Pole.

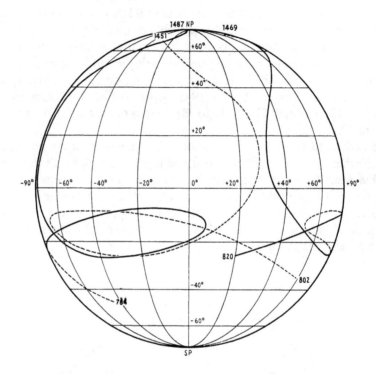

Figure 72. The total course of the exelegismos curves of the same eclipse series as in Figures 70 and 71. The dotted portions of the curves are to be imagined on the opposite side of the Earth.

Half of all solar eclipses (those at the northern node) belong to a Saros series which wanders from north to south over the Earth's surface; the other half (at the southern node) wander from south to north. They progress at eighteen-year intervals along a spiralling path around the Earth, so that each series gradually covers the whole Earth.

Figure 69 shows the progression of the series to which the eclipse of July 9, 1945 belongs, which was visible in its totality in Scandinavia and which, at its recurrence on August 11, 1999, will be visible as a total eclipse in northwestern and central Europe. It takes place at the northern node, and accordingly first appeared in the Arctic on January 4, 1639. Its eighteen-yearly recurrences waxed through fourteen partial eclipses, the last of which took place in 1873. The first totality followed on June 6, 1891, ushering in the series of total eclipses of June 17, 1909; June 29, 1927; July 9, 1945; July 20, 1963; July 31, 1981; August 11, 1999, and so on. The map in Figure 69 is centred at the North Pole showing only the totality zones. As the curves shift towards the south, a clear separation into

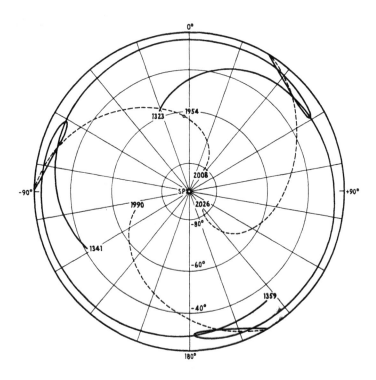

Figure 73. Exelegismos curves for another eclipse series, which also wanders from south to north. The South Pole is at the centre of the map.

three groups can be observed. The eclipses of each such group are separated by 54 (3 × 18) years. This triple Saros period was also known in ancient times, and was called the *exelegismos*. As the Saros comprises 18 years, $10\frac{1}{3}$ (or $11\frac{1}{3}$) days, the eclipse recurs in each instance about 8 hours later, and therefore, as a result of the Earth's rotation, appears displaced by about a third of the Earth's circumference. Not until 3 × 18 years have elapsed does the eclipse recur roughly at the same time of day, and the region of visibility is near that of the eclipse 54 years earlier.

Figure 70 shows the latter half of the totality zones of the eclipse series which wandered from the Antarctic to the Arctic in the period from January 3, 550 until April 8, 1902, and is since extinguished. Its totality lasted from 766 to 1523.

The principle which gives the pictures of these movements their so individual character can be clarified by connecting the mid-points of all the single curves within a group. These 'exelegismos curves' or 'guiding lines' are indicated on Figures 69 and 70. They do not follow the terrestrial meridians in a straight line

105

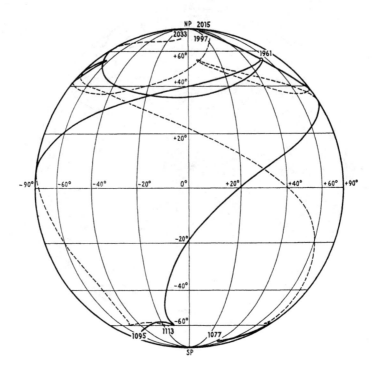

Figure 74. Exelegismos curves of the series which includes the eclipse of Feb 15, 1961. This series also wanders from south to north.

from Pole to Pole, but wind along in curves and loops of the most various kinds. The following illustrations give a few examples of the forms which arise in this

Figures 71 and 72 show the guiding lines of the eclipse series of Figure 70 from two further viewpoints.

Figure 73 shows those of the eclipse series, one of whose totality zones moves across northern Siberia on July 22, 1990. Its first appearance was in the Antarctic in 1179; from 1323 to 1810 it was annular; in 1828 and 1846 it was hybrid;* from 1864 to 2044 it is total; on May 3, 2495 it will disappear at the North Pole. Its total life-span is 1316 years.

Figures 74 and 75 provide two views of the guiding-lines of the totality zones of the series to which the total eclipse belongs, which was visible in the states of Washington and Montana, and in Manitoba, Canada on February 26, 1979. The totality and annularity of this series lie between August 11, 1059 and

*If an eclipse is partly annular and partly total it is called hybrid.

106

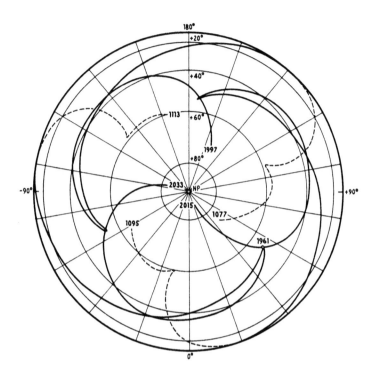

Figure 75. Exelegismos curves from the same series as in Figure 74, seen from the North Pole.

March 30, 2033; its first partial emergence near the South Pole was on May 27, 933; its last appearance will be in 2195 near the North Pole. Its life-span is 1262 years.

Figures 76 and 77 show the guiding-lines of the series which appears in the eclipses of April 29, 1976 (partially visible in southern Europe), and May 10, 1994 (visible in southern and eastern United States). The life span of this series is 1280 years, extending from September 9, 1002, where it first appeared near the South Pole, to November 1, 2282 when it will be visible for the last time in the Arctic. From May 16, 1417 to July 25, 2120 it is central.

The irregular movement of the guiding-lines from Pole to Pole is a result of the fact that the intervals of 54 years and 33 days between the individual members shift each time by a month in the seasons. Each exelegismos series, therefore, contains eclipses which take place successively at spring, summer, autumn and winter New Moons. The rising and sinking of the annual positions of the Sun and the New Moon have an accelerating or retarding effect on the

107

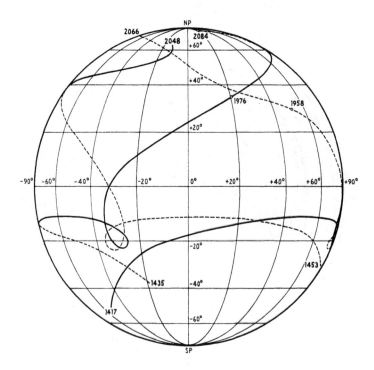

Figure 76. Example of exelegismos curves of another series also wandering northwards.

shift of the eclipses along the meridian. The additional lateral divergences arise through fluctuations in the Saros, due to the apsidal month and other irregularities.

Lunar eclipse series

Lunar eclipse series also wander over the Earth's surface. But here we find a polarity to the solar eclipses. A lunar eclipse is always visible over a whole hemisphere. The mid-point of the eclipse is always a point in the vicinity of the equator, whose geographical latitude is equal to the declination of the Moon. This mid-point wanders, during the exelegismos series, in waves or loops along the equatorial zone of the Earth. The predominant movement is west-east, in contrast to the solar curves, which have a predominantly north-south movement.

Let us consider two examples. The lunar eclipse series whose guiding-lines are shown in Figure 78 began on May 18, 1296 and will end on August 19, 2035; it occurs at the southern node and lasts 739 years. In the present decades the eclipses

108

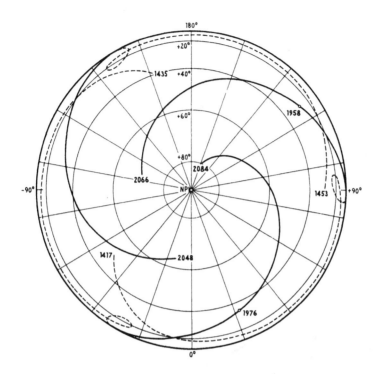

Figure 77. Exelegismos curves of the same series as in Figure 76, seen from the North Pole.

belonging to this series are partial and take place on July 17, 1981, and July 28, 1999.

Figure 79 shows the progression of the mid-point of an eclipse series which began on May 31, 1379 and will end on July 28, 2064; it has a life-span of 685 years. This Saros series takes place at the northern node. Its members include the partial eclipses of June 4, 1974; June 15, 1992; and June 26, 2010.

The curves of the first example are almost completely separated, and form closed loops; those of the second example embrace the whole circumference of the Earth more than once, and overlap strongly.

In earlier times it was the abnormal element of an eclipse which was experienced most strongly. The harmony of the cosmos and the course of everyday life on the Earth were disturbed, and seemed to falter. The myths spoke of a dragon or a wolf which devoured the Sun during an eclipse. A remnant of this imagination lives on in the designation of the lunar nodes as the 'dragon-points', whereby the

109

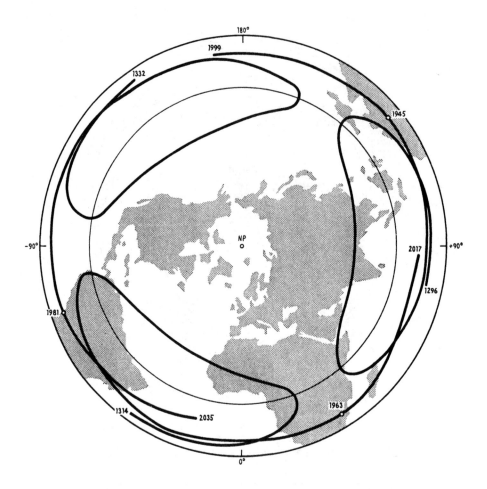

Figure 78. The exelegismos curves of a lunar eclipse series which show loop forms.

northern node was often represented as the dragon's head, and the southern node as the dragon's tail. For this reason the nodal month is still referred to as the draconitic month.

Yet the ordered regularity with which the eclipses, in all their complexity and variability, are distributed over the surface of the Earth, does not appear evil or obscure, but rather lucid and harmonious. All these interrelationships between Sun, Moon and Earth show clearly that no single one of these celestial bodies follows its own independent and arbitrary motion. The interaction of the most varied motions bears the mark of a unified, wisdom-filled character. It seems

110

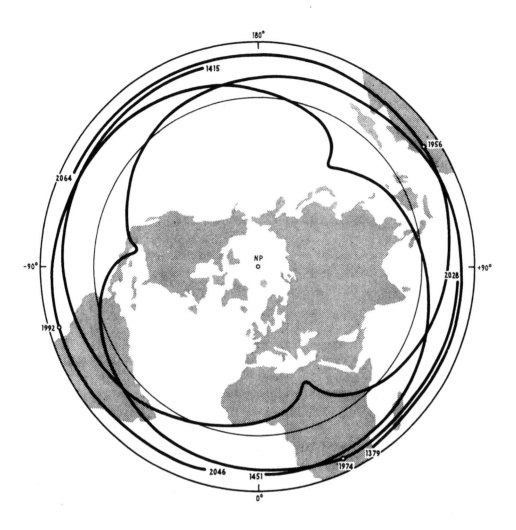

Figure 79. Another lunar eclipse series, this time showing wavy forms.

remarkable that the apparent size of the discs of Sun and Moon should be almost exactly identical. Thereby the Moon's shadow only just spans the distance to the Earth, its tip just touching the latter's surface. In the case of a solar eclipse when the Moon is near its apogee, the tip of the shadow ends somewhat above the Earth's surface: in the centrality belt an annular eclipse can be seen. Here too, we see how finely balanced the relationships are, for if the Moon were only one Earth-diameter further from the Earth (a factor of only a thirtieth of its total distance) it could only bring about annular eclipses of the Sun.

The planets

14. Phenomena common to Venus and Mercury

Like the Moon in the course of its monthly revolution, the individual planets display a cyclic repetition of their movements, following various characteristic rhythms. Moreover, we find an essential difference in this respect between Venus and Mercury on the one hand, and the remaining planets Mars, Jupiter and Saturn on the other. The outermost planets Uranus, Neptune and Pluto, which are invisible to the unaided eye, form an 'annex' to the latter group.

Venus and Mercury are most strongly characterized by the periodic alternation of their appearance as *morning* and *evening stars*. They are always relatively near the Sun, and remain within the same general field of observation as the latter. Mercury attains a maximum elongation from the Sun of 28° to the east or the west. The greatest possible elongation for Venus is about 48°. Even when the three luminaries Sun, Mercury and Venus reach their widest mutual separation, they always remain within the same celestial quadrant, and are in closely neighbouring constellations of the zodiac. From this aspect, Mercury and Venus are pre-eminently daytime planets. On a very clear day Venus, if at her brightest, can even be seen in the daytime. This is often the case in the dry atmosphere of deserts.

The actual time of observation for Venus and Mercury is limited to the evening and morning hours. When the planets stand in eastern elongation, they set after the Sun and are visible as evening star. They also rise after the Sun in the morning. Conversely, when Venus and Mercury are west of the Sun, they precede the latter in its daily course, and have already set when night begins. They can then not be seen in the evening sky, but are visible before sunrise in the eastern sky. Venus and Mercury can never reach quadrature with the Sun, to say nothing of opposition; therefore they can never be above the horizon at midnight.

The times of visibility as morning or evening star alternate with periods of invisibility. The planets' disappearance into the twilight at the end of an evening

Planet's position relative to the Sun	Planet's visibility	Planet's position relative to the Earth
Superior conjunction	Invisible	Apogee (behind Sun)
Heliacal rising	Becomes visible as evening star	
Greatest eastern elongation	Approximate time of longest visibility	
Heliacal setting	Becomes invisible	
Inferior conjunction	Invisible	Perigee (in front of Sun)
Heliacal rising	Becomes visible as morning star	
Greatest western elongation	Approximate time of longest visibility	
Heliacal setting	Becomes invisible	
Superior conjunction	Invisible	Apogee (behind Sun)

star period is known as the *heliacal setting*. Venus and Mercury then move into close proximity to the Sun, until they reach conjunction with the latter. Thereupon they cross over from the one side of the Sun to the other, changing from evening to morning star. Their first appearance in the morning sky is the *heliacal rising*. Two passages in opposite directions through non-visibility thus correspond to the two phases of visibility in the morning and in the evening. When they change from morning to evening star they wander from west to east past the Sun. This passage is then known as a *superior conjunction*. Conversely at the change from the evening to the morning sky a so-called *inferior conjunction* takes place, when the planets pass from east to west.

The meaning of the terms superior and inferior conjunctions becomes clear when we consider not only the planets' elongation, which can easily be followed in the sky, but also their changing distances from the Earth. We then find that Venus and Mercury are farthest from the Earth at the time of their superior conjunctions, when they stand beyond the Sun. During their inferior conjunctions they are nearer the Earth, standing between it and the Sun. In the former case, these planets are in their apogee, and turned, so to speak, towards the 'superior' cosmic reaches, while in the latter they are in perigee, and are turned to face the 'inferior' realm of the Earth. Their changing spatial relationship to the Sun is accompanied by a change of phase on the planets' discs, which can be observed with the telescope. We shall speak in greater detail about this in Chapters 16 and 17.

The complete cycle of Venus' and Mercury's phenomena leads through all the various stages until they return to the same aspect with the Sun. We define this as a synodic period. The alternation of evening and morning star periods takes place in different rhythms for the two planets. The most important stages, and the placements which they represent, are given in the table on page 113.

As can be seen from this outline of the stages of a synodic revolution, Venus and Mercury circle round the Sun. The planes of these orbits roughly coincide with the plane of the ecliptic. Seen from the Earth, they appear as flattened, elongated ellipses (Figures 83 and 98).

As the Sun wanders through the zodiac, the centre of the orbits of Venus and Mercury is continually shifted along the ecliptic. The orbital curves of the two planets therefore arise through the interplay of two movements: the planets' revolution round the Sun, and the simultaneous annual progression through the zodiac. The result is a cycloid (see Figures 84, 99 and 102).

15. The rhythms of Venus

The synodic period

Venus appears alternately in the evening and in the morning sky. At the times of its maximum brightness it makes a striking, and yet alluring impression in the twilight sky. It easily outshines all other planets and fixed stars.

During an evening star period, which always follows upon a superior conjunction, Venus can at first be seen only briefly in the twilight just over the western horizon. Its first appearance, the heliacal rising as evening star, takes place about 36 to 40 days after the superior conjunction, when Venus has attained about 10° elongation from the Sun. The duration of visibility increases very slowly. Only after six months is the greatest eastern elongation (46° to 48°) attained, and the longest visibility, during which Venus can be seen in the western sky up to four hours after sunset (Figure 80). At the same time, the planet progresses continually in the zodiac, which leads to a shift in its position over the western horizon, and the point where it sets. If the evening star period falls in the autumn, Venus' setting points, like those of the Sun, progress towards the south-west. If it falls in the spring, they will shift towards the north-west, as the planet is then moving in the ascending part of the zodiac.

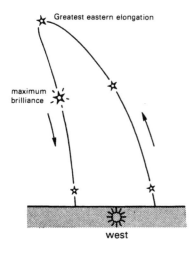

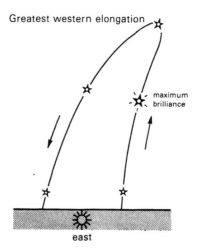

Figure 80. The rising and sinking of Venus during an evening star period in the west.

Figure 81. The rising and sinking of Venus during a morning star period in the east.

After the greatest eastern elongation has been attained, the separation between Venus and Sun decreases relatively rapidly, although the planet at first gains in brightness, reaching a maximum about 36 days after the greatest elongation. Thereupon, the evening star period rapidly nears its conclusion. A mere three weeks later the heliacal setting follows, when Venus again becomes invisible, with an elongation of 10° from the Sun. After a further twelve days it passes the latter in inferior conjunction.

The following morning star period forms, in its sequence, a mirror image to the course of events which we have just described. The maximum brightness is reached about 36 days after the inferior conjunction. The greatest western elongation (46° to 48°) with the longest visibility follows after a further 36 days. For the next six months the brightness and length of visibility slowly decrease, the latter sinking to less than an hour. About five weeks before the superior conjunction Venus has its heliacal setting (Figure 81).

Venus' invisibility lasts considerably longer around the superior than around the inferior conjunctions. At the latter, the transition from evening star to morning star visibility requires on the average about three weeks. The invisibility in passing from the morning to the evening sky lasts almost as many months.

The synodic period of Venus is about 1 year and 7 months. It can fluctuate between 577 and 592 days; the mean period is 584 days (1ª, 219ᵈ). Each evening and morning star period lasts about nine months.

An approximate unit of division for the Venus phenomena is given by the interval of 36 or 72 days. This can be seen immediately from the greatest elongation, the position of maximum brightness and the inferior conjunction, which follow each other at 36-day intervals. Between a superior conjunction and the previous or following greatest elongation lies the much longer interval of about 216 days (6 × 36), though this span of time can vary between 216 and 223 days.

Taken as a whole, therefore, a synodic Venus period is composed of 16 units of 36 days, or 8 units of 72 days. The subdivisions might be compared with 'months' in the 'year' of Venus phenomena. Moreover, as 36 days comprise about one tenth, and 72 days one fifth of a solar year, the Venus cycle can be clearly divided into units of a tenth or fifth of a year. The synodic revolution is then found to be, $\frac{16}{10}$ or $\frac{8}{5}$, that is, $1\frac{3}{5}$ years.

The most important positions of Venus are given below in sequence, with the corresponding conditions of visibility:

Venus' position relative to the Sun	*Average interval (days)*			*Visibility*
Superior conjunction				Invisible
	6 × 36		3 × 72	
Greatest eastern elongation				Longest visibility as evening star
	36			
			1 × 72	
Maximum brightness				
	36			
Inferior conjunction				Invisible
	36			
			1 × 72	
Maximum brightness				
	36			
Greatest western elongation				Longest visibility as morning star
	6 × 36		3 × 72	
Superior conjunction				Invisible
One synodic period	= 16 × 36 = 8 × 72			= 576 days (actually 584^d)

The periods of visibility, as well as the superior and inferior conjunctions of Venus are given in the Appendix (Table 6.5 and 6.4) for the years 1979 to 2010.

Owing to the length of the synodic Venus period, successive cycles of the Venus phenomena occur later in the year, and therefore in different parts of the zodiac. The evening and morning star periods accordingly fall at different times of the year. At the same time the conditions of visibility are affected. The least striking evening star periods are in the autumn, as for example in 1986, 1989 and 1995. In these months the zodiac is flattest with respect to the horizon at sunset; Venus therefore remains near the south-western horizon, despite its considerable elongation. If, however, the evening star period falls in the first half of the year, as for example in 1985, 1988 and 1993, very favourable conditions of visibility are given. The steep position of the western zodiac on spring evenings allows Venus to appear high above the horizon. Conversely, Venus is most impressive as a morning star when it reaches its greatest elongation in the second half of the year (for example in 1983, 1988 and 1991). The morning star periods which fall in the springtime are relatively unfavourable for observation (for example 1986, 1989, and 1992). In the southern hemisphere, the best visibility is always when unfavourable in the north.

Five successive synodic Venus periods are quite different in character. These take place over a period of almost exactly eight years. To be more precise, 5 synodic periods last 2919.60 days, while 8 years last 2921.94 days, giving a difference of only 2.34 days. Accordingly, all Venus phenomena occur 2 days, 8 hours earlier after the completion of an eight-year cycle.

Every sixth Venus period is therefore remarkably similar; the various stages are repeated in almost the same form. With the eight-year period a kind of Venus-calendar is given, in which the total succession of phenomena can easily be surveyed for longer periods of time.

The sidereal period

Thus far we have only considered the relation of Venus to the Sun. All of the motions we have discussed, however, take place against the background of the zodiac, which changes with the Sun's yearly movement. A retrograde motion does not at first strike our attention. More exact observation is required to show that Venus moves for a time backwards through the zodiac. It begins on average fourteen days after the maximum brilliance in the morning sky. The retrograde movement takes about 41 days, and extends over about 17° in the zodiac. As the inferior conjunction falls during this time, with the three-week long invisibility of the planet, and as the fixed stars can seldom be seen in the twilit sky, the retrograde movement is hardly noticeable for ordinary observation.

During each synodic Venus period, therefore, a loop or open S-shaped curve is traced by the planet in its movement at the time of the inferior conjunction. Two such Venus loops are separated by an interval of 1 year 7 months ($1\frac{3}{5}$ years). This scale can at once be found in the distribution of the individual Venus loops in the zodiac. As Venus always remains more or less in the neighbourhood of the Sun, it requires on the average an equal length of time to complete one revolution along the ecliptic, that is, one sidereal year. The $1\frac{3}{5}$ year synodic period likewise corresponds to $1\frac{3}{5}$ revolutions through the zodiac. Upon the completion of one of its loops, Venus wanders first through all twelve constellations, and then repeats three-fifths of this journey ($3 \times 72°$) before the following loop is begun. Figure 82 shows the movement of Venus in the zodiac for two successive loops.

The following loops are described in the zodiac at similar intervals. Five

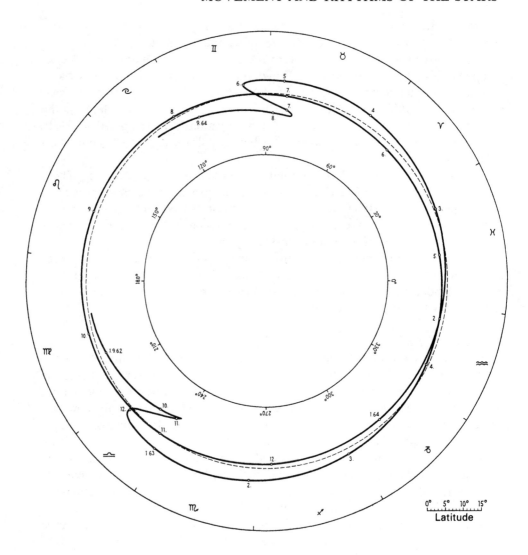

Figure 82. Successive Venus loops (1962–64).

successive loops are arranged symmetrically at roughly equal intervals of 72° (Figure 83). Each sixth loop nearly coincides with the first, as the planet completes five synodic periods, that is eight revolutions through the zodiac, in eight years $(5 \times \frac{8}{5})$. The movement during an eight-year cycle shows a strict pentagonal arrangement.

Alongside this cycle, the half-period of four years also has a special significance

118

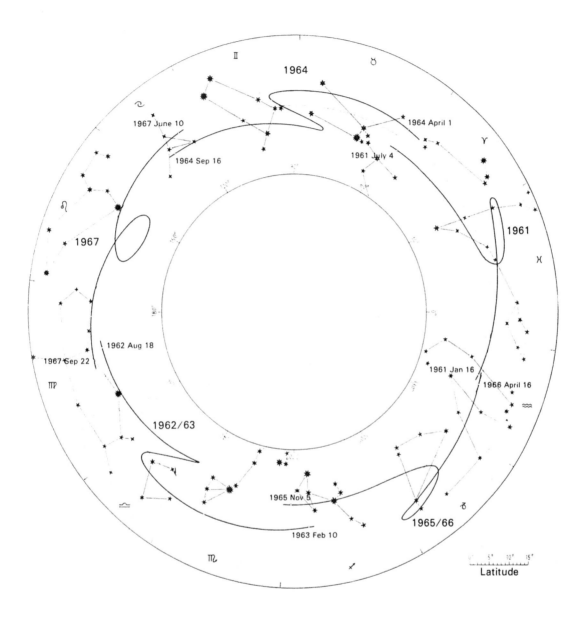

Figure 83. The loops of Venus 1961–67. Between two successive loops Venus wanders $1\frac{3}{5}$ times through the zodiac. The loops for each successive eight-year period are displaced by about 2°3 to the west (clockwise). The loop for 1975, therefore, is only about 2°3 west (direction of the Crab) of that for 1967, overlapping it strongly. The following loop (1977) takes place nearly 5° west (direction of Water-Bearer) of the loop for 1961.

119

for the course of Venus' rhythms. The whole eight-year cycle comprises five synodic periods, with five superior and five inferior conjunctions. After four years, therefore, exactly $2\frac{1}{2}$ synodic periods are completed, so that, if we begin with a superior conjunction, we can expect an inferior conjunction to take place four years later, almost to the day.

If we group together all the inferior and superior conjunctions occurring around a certain date (and consequently in a certain region of the zodiac) we obtain five series separated by 73 days, or 72°. In each series the superior and inferior conjunctions alternate at four-year intervals. The superior conjunction of August 23, 1987, for example, is followed by the inferior conjunction of August 22, 1991, the superior conjunction of August 20, 1995, the inferior conjunction of August 20, 1990, and so forth. This series occurs in the constellation of the Lion. Table 6.4 in the Appendix gives a longer survey of all the series of conjunctions between Venus and the Sun with respect to the four and eight-year rhythms.

The conjunctions are therefore restricted to five times of the year, separated by intervals of about 73 days. At present (1986) the conjunctions can only take place in the middle of January, at the beginning of April, in mid-June, at the end of August and at the beginning of November. Correspondingly, the meetings between Sun and Venus can only take place in 5 narrowly defined zones of the zodiac. These are presently located in the constellations Goat/Archer, Fishes, Bull, Lion and Scales/Virgin. In Figures 85 and 86 the five regions of the zodiac are shown for the conjunctions which took place in the period from 1952 to 1961. Successive superior as well as inferior conjunctions are separated by 144°, forming a regular pentagram. The sequence which the conjunctions follow during the eight-year period is given by the numbers 1 to 6 and 1' to 6'. Their positions can also be seen from Figure 84.

Let us now consider the positions of maximum brightness and greatest elongation during an eight-year cycle. Just as ten conjunctions (five superior and five inferior) are contained in such a period, so too ten positions of maximum brightness and ten greatest elongations are reached. These important stations of Venus' movement also stand in symmetrical relationship in time and space.

Figure 87 shows the distribution in the zodiac of Venus' positions of greatest elongation. The connecting lines give the temporal sequence, and form the intricate but closed configuration of a five-membered decagram. Figure 88 shows the corresponding picture for the positions of maximum brightness. Surprisingly, each position of the Sun corresponds to two Venus positions. In Figure 87, for example, the Sun is shown in the constellation of the Fishes on April 10. To this date belong the Venus elongation of April 12, 1956 (3E) as well as that of April 9, 1958 (4M), two years later. Similar relationships prevail for the other points in Figures 87 and 88. Temporal and spatial relationships are intricately interwoven in these figures.

A summary of the previous pictures leads to the following results: Within the

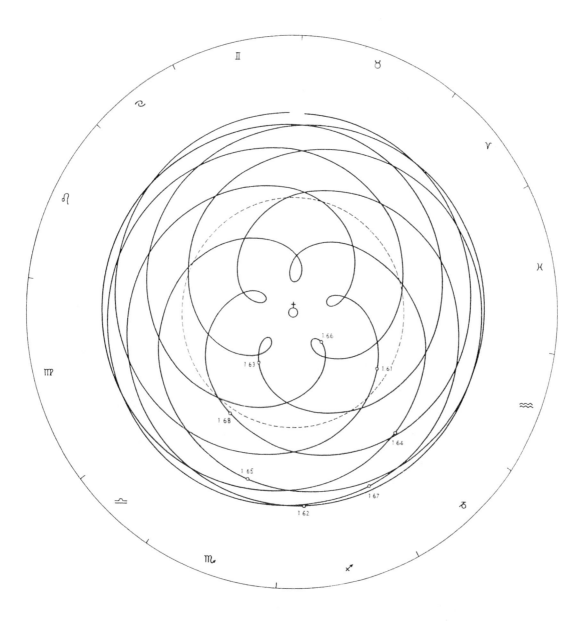

Figure 84. The geocentric Venus orbit 1960–68, corresponding to Figure 83. The small circles indicate the positions at the beginning of each year. The directions of the constellations are shown at the outer edge. Sun's orbit (– – – –).

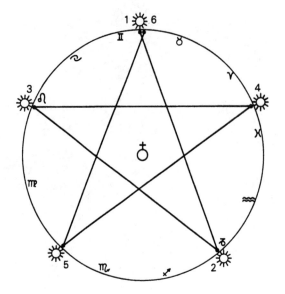

Figure 85. Distribution and sequence of the superior conjunctions of Venus and Sun in 1952–61.

1   1952 June 24
2   1954 Jan 30
3   1955 Sep 1
4   1957 April 14
5   1958 Nov 11
6   1960 June 22

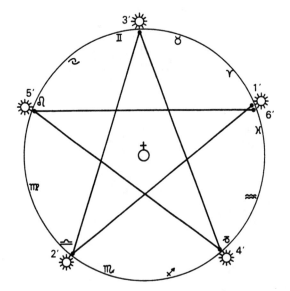

Figure 86. Distribution and sequence of the inferior conjunctions of Venus and Sun in 1952–61.

1'   1953 April 13
2'   1954 Nov 15
3'   1956 June 22
4'   1958 Jan 28
5'   1959 Sep 1
6'   1961 April 11

eight-year Venus cycle, at all conjunctions and all greatest elongations the Sun stands in five distinct positions in the zodiac (Figures 85, 86, 87). Midway between these points are the Sun's positions at the time of Venus' maximum brightness (Figure 88). Taken as a single, composite picture, these ten positions form a regular decagon.

122

Figure 87. Distribution and
sequence of the positions of
Venus and the Sun in 1952–61,
at the time of greatest
elongation as evening star (E)
and morning star (M).

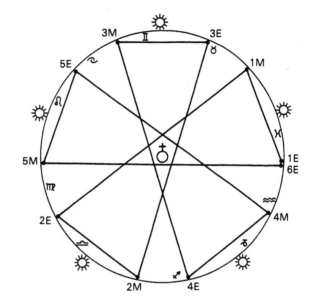

1E	1953 Jan 31
1M	1953 June 22
2E	1954 Sep 6
2M	1955 Jan 25
3E	1956 April 12
3M	1956 Sep 1
4E	1957 Nov 18
4M	1958 April 9
5E	1959 June 23
5M	1959 Nov 12
6E	1961 Jan 29

Figure 88. Distribution and
sequence of the positions of
Venus and the Sun in 1952–61,
at the time of maximum
brilliance as evening star (E)
and morning star (M).

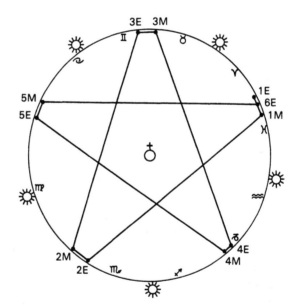

1E	1953 March 8
1M	1953 May 19
2E	1954 Oct 15
2M	1954 Dec 21
3E	1956 May 16
3M	1956 July 29
4E	1957 Dec 24
4M	1958 March 11
5E	1959 July 26
5M	1959 Oct 8
6E	1961 March 5

It may be remembered that we divided the single 'Venus year' (584 days) into subdivisions of 36 days, or tenths of a year. The eight-year period of Venus shows the same structure in its spatial form. The zodiac, which represents the annual path of the Sun in 365 days, and which is mathematically divided into 360°, appears here divided into 10 segments of 36° each.

The movement of Venus is dominated by the almost exact eight-year rhythm, at the end of which very nearly the same positions and spatial relationships are repeated. The difference from the true course is 2⅓ days; the progression in the sky, correspondingly, is 2°3.

The pentagon of conjunctions — and with it all other positions — is therefore displaced in clockwise direction over the course of an eight-year period by 2°3. After about 30 such segments have been described, an angle of 72° is attained: that is, the five corners progress each by a fifth of a circle in a period of roughly 8 × 30 = 240 years (in fact, 243 years), or 152 synodic periods. In the course of this slow progression, all Venus positions complete a full revolution around the zodiac in 1215 years (5 × 243) with 760 synodic periods.

16. The movement of Venus around the Sun

Seen through the telescope the surface of Venus shows slight differences in brightness. These however, are not permanent, and so do not permit a dependable calculation of the planet's period of rotation. Radar measurements, which have been carried out since 1964, indicate that Venus rotates over a period of 243 days in retrograde direction.

The most striking feature of the planet when seen telescopically is its change in phase, similar to that of the Moon. Moreover, the diameter of the planet's disc changes considerably along with the phases. These changes can be observed with a relatively low telescopic magnification. The changing shape of the illuminated portion of the planet in its various positions relative to the Sun bears witness to the fact that Venus does not radiate its own light, but mirrors the light of the Sun.

At the time of a superior conjunction Venus appears as a full, round disc with an apparent diameter of 10″, 'Full Venus'. Thereupon it appears to move eastward, away from the Sun; its illuminated portion decreases as its diameter increases. At the time of greatest elongation it has a half-moon form and a diameter of 25″. In the position of greatest brightness Venus has the form of a slender but large crescent (diameter nearly 40″). As it moves towards its inferior conjunction, the crescent becomes narrower and finally disappears at 'New Venus'. In this phase its diameter is greatest at 62″ (Figure 89). Between the inferior and the superior conjunction phase and size change in the reversed sequence: Venus first appears in the morning twilight as a large, slender crescent, decreasing in diameter while its phase waxes. At its greatest western elongation it is once again 'Half Venus', then slowly it moves to its heliacal setting at dawn, and on to 'Full Venus'.

124

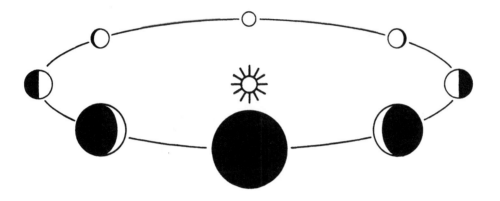

Figure 89. The change in phase and apparent size of Venus during one synodic period.

The duration of the full phase, from the greatest elongation as morning star through the superior conjunction to the greatest elongation as evening star, is three times as long as the new phase, which centres around the inferior conjunction.

In the course of Venus' phases we find some remarkable contrasts to the form and sequence of the lunar phases. The planet appears as Full Venus in the evening sky, wanes, and disappears as a crescent. This waning phase, however, shows the same forms as that of the waxing Moon. The waning Half Venus at greatest elongation, for example, has the form of the waxing Moon at first quarter. Correspondingly, the waxing Venus in the morning sky appears in the form of the waning Moon. It appears as a sickle and disappears as Full Venus.

While the Moon's diameter also changes, it is by no means in the same measure as that of Venus. Moreover, the Moon's variations have no connection whatever with the phases.

Further, the planet's maximum brightness does not coincide with its full phase, as its diameter is then smallest. Venus' positions of greatest brightness lie between greatest elongation and inferior conjunction, where the planet appears in crescent form.

From these changes in phase and in diameter, as well as from various further angular measurements, it can be concluded that Venus' distance from the Earth changes. In terms of astronomical units (the mean distance between Earth and Sun) Venus' distance at apogee, in superior conjunction, is 1.73, and at perigee, in inferior conjunction, 0.27. The mean distance is of course the same as that of the Sun from the Earth.

The following table summarizes these relationships:

Venus' position relative to the Sun	Phases and size		Distance in AU
Superior conjunction	Full Venus	○	1.73 Ag
Greatest eastern elongation	Eastern quadrature	◑	0.68
Maximum brightness		◖	0.46
Inferior conjunction	New Venus	●	0.27 Pg
Maximum brightness		◗	0.46
Greatest western elongation	Western quadrature	◑	0.68
Superior conjunction	Full Venus	○	1.73 Ag

Each evening star period begins with the Venus apogee (at superior conjunction). The evening star is therefore continually approaching the Earth. During a morning star period the opposite relationship prevails: the planet's distance from the Earth is steadily increasing. The speed of Venus' movement toward and away from the Earth also changes rhythmically, reaching its maximum at the times of longest visibility during greatest elongation.

Venus' positions relative to the Earth are affected in a twofold way by its rotation around the Sun: they display both a lateral movement along the ecliptic and a radial movement towards and away from the Earth. It is due to the latter movement that Venus is six times nearer the Earth in its loop at inferior conjunction than during superior conjunction. Out of the interaction of these two factors in Venus' movement arises the particularly beautiful picture of Figure 84. At its apogee Venus approaches nearer to the Earth than any other planet. Apart from the Moon it approaches closer to us than any other luminary visible to the unaided eye.

The revolution of Venus around the Sun can also be portrayed in another way. The differences in right ascension and declination of the two luminaries varies

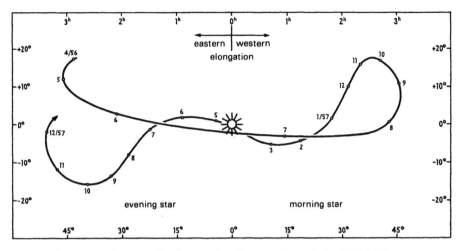

Figure 90. Venus relative to the Sun during the synodic period of 1956–57.

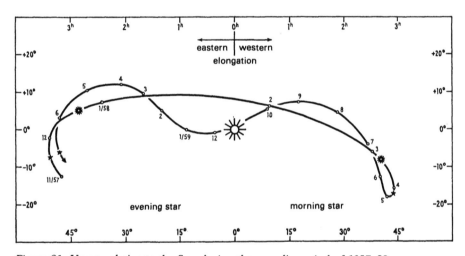

Figure 91. Venus relative to the Sun during the synodic period of 1957–59.

daily. These differences can be plotted in diagrammatic form from day to day, with the Sun as a fixed point. The line connecting these points forms a curve, as in Figure 90. Left of the Sun is Venus in eastern elongation as evening star; to the right, in western elongation as morning star; its ecliptical longitude is in the former case greater, in the latter case less than that of the Sun. If the planet is higher than the Sun, it has a higher ecliptical latitude; if it is lower, the latitude is also lower. The small numerals give the positions of Venus for the first of each month. Figure 91 begins at the left with November 1, 1957. On November 15 the

127

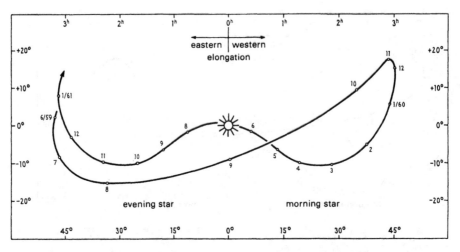

Figure 92. Venus relative to the Sun during the synodic period of 1959–61.

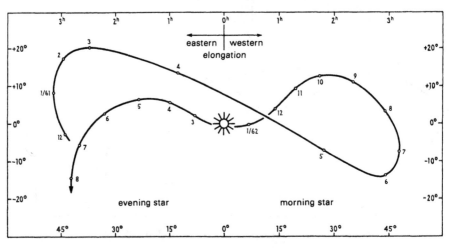

Figure 93. Venus relative to the Sun during the synodic period of 1961–62.

planet attained its greatest eastern elongation (small star) with 47½°. Thirty-nine days later, on December 24 (large star) it reached its maximum brightness. After a further thirty-five days the inferior conjunction with the Sun took place (January 28, 1958). Thereupon the morning star period began, in which Venus, on March 4, after a further thirty-five days, had once again attained its maximum brightness (large star). Thirty-six days later on April 9 followed the greatest western elong-

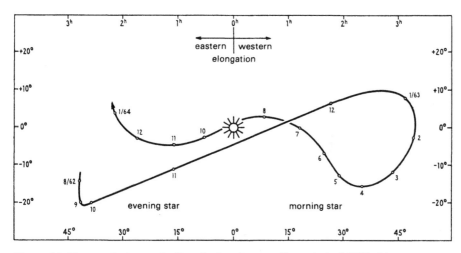

Figure 94. Venus relative to the Sun during the synodic period of 1962–64.

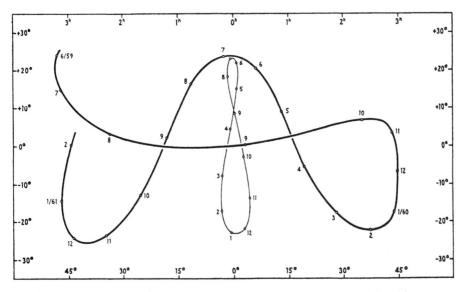

Figure 95. The curve corresponding to Figure 92 if the simultaneous change in the midday position of the Sun is taken into account (June 1959 to Feb 1961).

ation (46½°, small star). The superior conjunction with the Sun took place 216 days later on November 11; the planet then became an evening star. With the greatest eastern elongation (45½°) on June 23, 1959 (after 224 days) the circuit of this synodic period was closed. In the diagram the faster movement of Venus at the time of inferior conjunction (perigee), as compared with the slower movement around the superior conjunction, can be clearly seen.

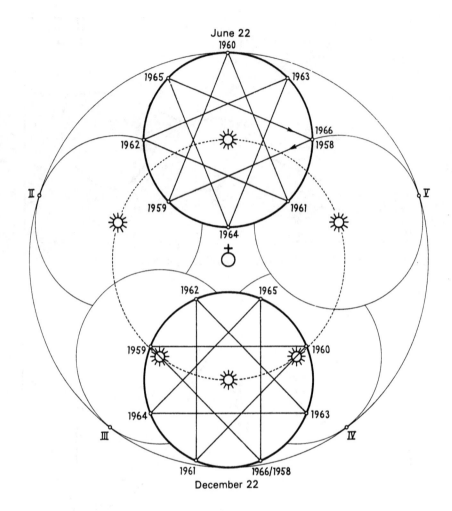

Figure 96. Octagram of the positions of Venus from 1958 to 1966: above for June 22, below for Dec 22.

Five characteristic curves can be drawn in this manner for the eight-year Venus rhythm; the sixth would be almost identical with the first (Figures 90–94).

In the diagrams the Sun is assumed to be at rest. We have seen in Chapter 7, however, that the positions of the Sun at noon local time describe a lemniscate in the course of a year. If this is taken into account, a curve emerges such as that in Figure 95, which is transformation of Figure 92. If Venus and the Sun could be photographed daily on a single plate at noon local time over the course of a synodic Venus period, this figure would result.

It can similarly be asked, what positions does Venus occupy on the same date in successive years? In accordance with Venus' eight-year cycle, we obtain only eight different positions for a given date which, however, follow a regular pattern. Figure 96 is drawn geocentrically, so that not only the planet's position relative to the Sun, but also its distance from the Earth (in the centre) is represented.

If the positions for eight successive years (for instance, 1958–66) are connected, a regular octagram is obtained. The four-year Venus period also comes to expression here, as the planet stands at opposite corners of the octagram at four-year intervals. It can be seen that the superior conjunction of June 22, 1960 was preceded four years earlier (1956) and followed four years later (1964) by an inferior conjunction in the same part of the zodiac, at which Venus stood at perigee on the line between Earth and Sun (see Chapter 15).

The same observation can be carried out for any other date of the year, whereby each day is characterized by a different position of the Sun in the zodiac and therefore by a different situation of Venus' orbit. In each case the octagram

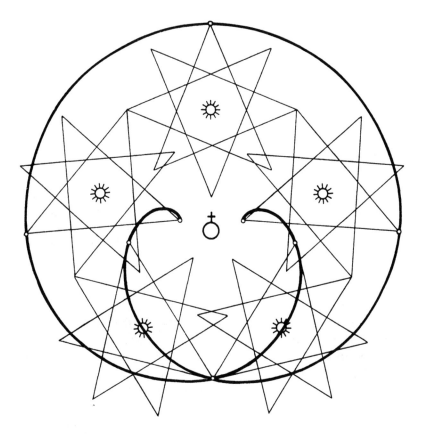

Figure 97. The rotation of the octagram of the positions of Venus during one synodic revolution (1959–61).

lies differently. Figure 97 illustrates only those positions of the octagram in which it returns to the same orientation to the Earth. Altogether there are five such positions, following at intervals of roughly 73 days, and at a distance of 72° (indicated in Figure 96 by the succession of circles I, II, III, IV and V).

In Figure 96 the Venus octagram for December 22 is shown directly opposite the upper eight-pointed star. This opposite orientation to the Earth also returns five times, forming a second set of five, midway between the first set.

The continuous transition through all further stages may be followed if we imagine that the passage of the octagram from position I to position II (72°) is accompanied by a rotation of the octagram by 45°, the angle between two adjacent tips. (The combination of corresponding portions of the two movements gives a simple and elegant construction of the epicyclical Venus movement).

In this way it is possible, by determining the position of the octagram for a sufficient number of days in the year, to construct the looping movement of Venus around the Earth. This is indicated in Figure 97, beginning with the superior conjunction (June 22, 1960) and proceeding forwards and backwards to the following and the preceding inferior conjunction. The path touches each octagram at the next point, thus slowly moving inwards.

A whole series of further stages in the Venus cycle (for instance, greatest elongation and maximum brightness) can be derived from these drawings.

17. The rhythms of Mercury

Of all the planets which can be seen with the naked eye Mercury is most seldom visible. Although it has several morning and evening star periods each year, which total a few weeks, the duration of its visibility to the unaided eye in temperate latitudes is, even under the best conditions, only about fifteen to eighteen hours annually.

The synodic period
Like Venus, Mercury is decidedly a 'daytime planet': that is, it stands for many hours above the horizon in the vicinity of the Sun, whose light totally outshines it. Mercury's position relative to the Sun changes in regular rhythms. If it stands to the east (left) of the Sun, it will set after the latter, and is therefore an evening star. In western (right) elongation it sets before the Sun, but is visible before the latter in the morning sky (corresponding to Figure 80 and 81).

As a result of the small maximum eastern and western elongations of a mean 23° (18° to 28°) the possibility of observing Mercury with the unaided eye is limited. The planet usually remains too near the horizon in the morning or evening sky

132

and is veiled by the atmospheric haze. There are, nonetheless, favourable periods of visibility, during which Mercury can be seen in the evening or morning sky for about an hour, over a period of around a week. It then shines with roughly the brightness of Sirius near the rising or setting point of the Sun, 10° to 15° above the horizon. As for Venus, the best conditions of visibility are for the evening star in springtime, and for the morning star in autumn. At these seasons the zodiac rises or sets most steeply at the corresponding times of day, so that Mercury stands relatively high in the twilit sky at the times of its greatest elongation.

Mercury's movements are in all respects similar to those of Venus, with the difference that they take place more rapidly.

After a superior conjunction, Mercury moves at first slowly away from the Sun. Then, however, after attaining its best evening visibility, it disappears quickly into the neighbourhood of the Sun. Conversely, in the following morning star period, it soon reaches its period of optimum visibility, after which it fades slowly into the morning twilight. Correspondingly, the period of non-visibility before and after an inferior conjunction — immediately preceding a morning star period — is much shorter than the non-visibility surrounding a superior conjunction with the Sun.

The positions of favourable visibility are much more strongly confined to the proximity of the greatest elongations than is the case with Venus. They accordingly fall nearer in time to the inferior than to the superior conjunctions; the proportion being roughly 3:2.

Mercury's synodic period, or the cyclic return of the individual stages in its movement through the evening and morning star periods, lasts about four months. The average duration is 116 days (more exactly, 115.88 days). It can, however, vary between 104 and 132 days for individual synods. The phenomena such as the conjunctions and the greatest elongations from the Sun take place at distinct intervals within the greater period. On the whole, we find a rough division into intervals which are a multiple of 12 days, as shown in the table overleaf.

The four-monthly synodic Mercury period is repeated three times in a year, so that three evening and three morning star periods take place annually, and in general three superior and three inferior conjunctions, and six greatest elongations.

Three synodic Mercury periods do not, however, embrace a full year, but are eighteen days shorter ($3 \times 116 = 348$). Accordingly, all the Mercury phenomena take place about $2\frac{1}{2}$ weeks earlier from year to year.

These seasonal progressions are immediately evident when we glance over a table of the dates of the Mercury phenomena for a longer period of time. (Compare the Mercury dates 1971–2010 on Table 6.3 of one Appendix). The inmost planet accompanies, more or less faithfully, the Sun in its course through the zodiac. It thereby passes through the rising portion of the zodiac, roughly from the Archer to the Bull, in the first half of the year, and through the descending portion from the Twins to the Scorpion in the second half of the year.

133

Mercury's position relative to the Sun	Average interval (days)	Visibility
Superior conjunction	} $36 = 3 \times 12$	Invisible
Greatest eastern elongation	} $22 \simeq 2 \times 12$	Evening star
Inferior conjunction	} $22 \simeq 2 \times 12$	Invisible
Greatest western elongation	} $36 = 3 \times 12$	Morning star
Superior conjunction		Invisible
One synodic period	116 days $\simeq 10 \times 12$	

The eighteen-day difference between the three Mercury synodic periods and the year accumulates over the course of seven years to about 123 days, which allows an additional synodic period. After seven solar years and twenty-two Mercury synodic periods, therefore, all the planet's phenomena are repeated $7\frac{1}{2}$ days earlier. This is a first greater comprehensive rhythm of Mercury.

Not until the passage of 46 years, almost seven times as long, is a cycle of altogether 145 synodic Mercury periods completed, after which all Mercury phenomena return with a delay of one day. This 46-year planetary period of Mercury was already known in ancient times to Hipparchus and Ptolemy. It is particularly useful for determining future positions of the planet. For this purpose the 46-year period was used as the basis for the medieval planetary ephemerides, the Alphonsine Tables (1252 in Spain), in which the positions of Mercury were given for a span of 46 years.

The differences which always remain, even after the longest periods of time, show that a simple period in terms of whole years is never adequate to define the course of a planet's movement. It is one of the most characteristic features of all the wandering luminaries, that their movements never repeat themselves mechanically in all details, but remain in a state of ever-changing flux, in which new variations continually arise.

The sidereal period

The direct observation of Mercury's movement with respect to the zodiac is even more difficult than for Venus, as the former planet never passes out of the region of bright twilight. As we have already seen, Mercury, on account of its dependence

134

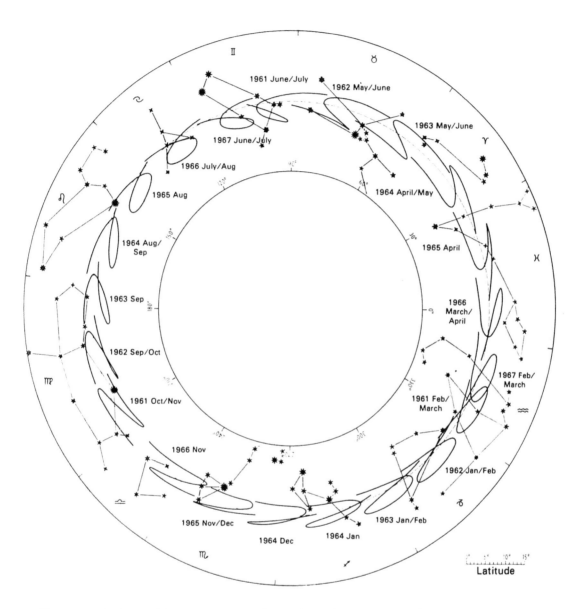

Figure 98. The loops of Mercury 1961–67. The portions of the orbit which lie between successive loops (about 115° in length) have been omitted. The loops for each seven-year cycle are displaced by about 7° to the west (clockwise). The loop for October 1975, for example, takes place about 14° west (direction of Lion) of the loop for Oct/Nov 1961, overlapping only slightly. Its form more closely resembles that of the loop for Sep/Nov 1962.

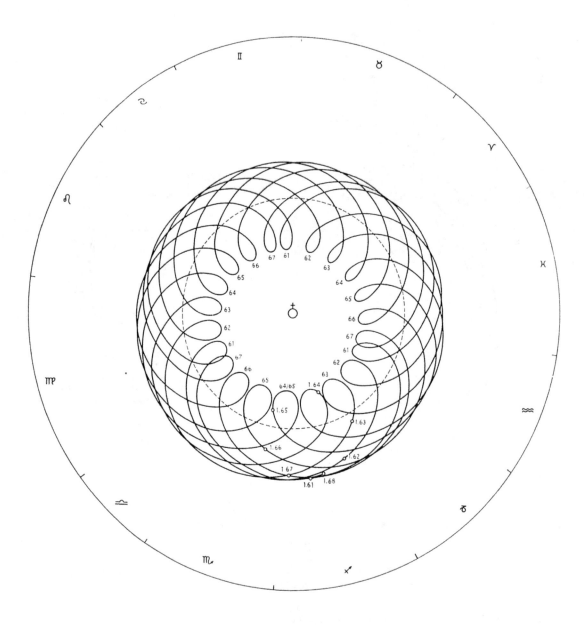

Figure 99. The geocentric orbit of Mercury 1961–67, corresponding to Figure 98. The numbers indicate the years belonging to the different parts of the orbit. The directions of the constellations are shown on the outer edge. The dashed circle indicates the Sun's orbit.

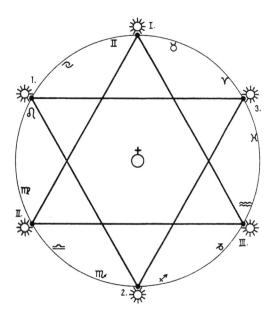

Figure 100. Hexagram of the most important positions of Mercury during one year. The superior conjunctions (I, II, III) are drawn schematically in the same manner as the inferior conjunctions (1, 2, 3).

on the Sun, completes a full revolution through the zodiac in a year. A retrograde movement with the formation of a loop is still less evident than for Venus. This backwards motion begins ten to fourteen days before the inferior conjunction and ends roughly as many days after; altogether it lasts nineteen to twenty-five days. The retrograde portion of its orbit varies considerably in size. For the smallest loops, in the constellation of the Bull, it is 8°.5; in the opposite region of the zodiac, in the Scorpion, the loops are nearly twice as long, tracing an arc of up to 16°.5.

In each synodic period Mercury describes a loop, so that altogether three loops take place each year. Mercury's movement from west to east is accordingly interrupted each four months by a loop. The interval of about a third of a year between successive inferior conjunctions corresponds to a progression of the loops by about a third of the zodiac. It follows that in the course of a year the three Mercury loops are distributed over the zodiac at roughly equal intervals of 120°. As three Mercury synodic periods are 18 days shorter than a year, the loops of successive years are grouped relatively close together, but do not overlap (see Figure 98).

The positions of Mercury's inferior conjunctions with the Sun, which lie midway in the loops, form a roughly equilateral but slightly open triangle in the zodiac.

137

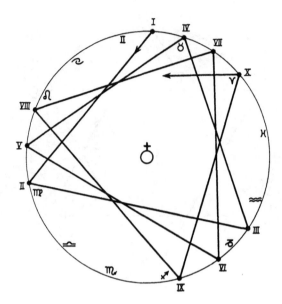

Figure 101. The rotation of the trigon of the superior conjunctions of Mercury, opposite to the direction of the zodiac.

I	1958 June 18	VI	1960 Jan 26
II	1958 Oct 5	VII	1960 May 17
III	1959 Feb 14	VIII	1960 Aug 31
IV	1959 June 3	IX	1961 Jan 5
V	1959 Sep 17	X	1961 May 1

This also applies to the superior conjunctions, though they are displaced by about 60° against the inferior ones. The most important stations in Mercury's movement thus form two interpenetrating triangles in the zodiac each year (Figure 100). The triangle described by Mercury in one year corresponds to the pentagram which Venus describes in eight years, over the course of five synodic periods. The hexagram showing both inferior and superior conjunctions of Mercury also corresponds to the Venus pentagram, as Venus' superior and inferior conjunctions take place in the same part of the zodiac (see Figures 85 and 86).

Naturally, the rhythms of Mercury are very much less regular than those of Venus, so that the following indications are average values which can vary considerably.

The trigon of the superior conjunctions — and similarly that of the inferior conjunctions — is not a closed figure, but regresses in the zodiac by about 18° each year (Figure 101). In other words, the points of the hexagram are displaced by slightly more than one constellation in two years. As a result of the movement of the points, that is, of the positions of the conjunctions, the superior conjunctions take place at the original positions of the inferior conjunctions after the course of 9½ synodic Mercury periods, or three years, with a delay of only five days, or 5°.

After the completion of the seven-year cycle with its 22 synodic periods, each corner of the two trigons has moved backwards through the zodiac, almost to the original position of the adjacent corner; the Mercury phenomena, however, take place seven days (or 7°) earlier. Not until 46 years have elapsed with 145 synodic periods do the events correspond in time and space, with a deviation of only one day, or 1°.

18. Mercury's movement around the Sun

When viewed through the telescope, Mercury, like Venus, shows a change of phase. At the same time, the planet's apparent diameter changes. During a superior conjunction Mercury is 'full', with a diameter of as little as 4″.5. As the planet moves eastward, away from the Sun, its diameter gradually increases, and the phase becomes gibbous. 'Half-Mercury' is attained at greatest eastern elongation. It now begins to approach the Sun, and at the same time increases in size, and passes over into a crescent phase. At the inferior conjunction Mercury attains its greatest apparent diameter of 13″. It is then 'New Mercury'. Its diameter is about a fifth that of Venus at inferior conjunction. In the subsequent morning star period the phases and the diameter of Mercury change in opposite sequence to that in the evening sky. The planet, emerging as the large 'New Mercury', is 'half' at greatest elongation, and disappears as the small 'Full Mercury' shortly before the superior conjunction. Like Venus, Mercury displays a contrast to the Moon in the sequence of its phases.

Since the disc of Mercury is 'half' at greatest elongation, it follows that the duration of the full phase, from one Mercury quarter over the superior conjunction to the other quarter, is longer than that of the new phase. Again, this shows a similarity to Venus.

The changes in phase and diameter, together with further angular measurements, point clearly to a varying distance between the Earth and Mercury. In terms of astronomical units (Earth-Sun distance) Mercury's distance from the Earth varies between 1.39 AU at superior conjunction and 0.61 AU at inferior conjunction. Owing to Mercury's uneven motion, these values can fluctuate by as much as 0.10 AU. The mean distance is, of course, equal to the distance between the Sun and the Earth. It must be borne in mind, however, that Mercury, like Venus, spends more time beyond the Sun's orbit than within it: the relationship is roughly 3:2.

As the evening star periods begin with Mercury's apogee at a superior conjunction, and end with its perigee at the inferior conjunction, it is evident that Mercury is continually approaching the Earth during this time.

Conversely, in the morning star period, the planet's distance from the Earth continually increases. The speed at which this distance changes is not constant, but follows a rhythmic fluctuation. It is slowest around the time of the conjunctions, and fastest at the times of greatest elongation.

Mercury's position relative to the Sun	*Phases and size*		*Distance in AU*
Superior conjunction	Full Mercury	○	1.39 Ag
Greatest eastern elongation	Eastern quadrature	◑	0.92
Inferior conjunction	New Mercury	●	0.61 Pg
Greatest western elongation	Western quadrature	◐	0.92
Superior conjunction	Full Mercury	○	1.39 Ag

From the foregoing it is evident that Mercury circles around the Sun like Venus. Its orbit is, however, closer to the Sun than that of Venus. Its greatest elongation is on average 23°, roughly half that of Venus. Mercury therefore cannot approach so near to the Earth as the latter.

Out of the interplay between Mercury's spherical movement in the zodiac and its radial movement towards and away from the Earth, arises the planet's geocentric orbit. Mercury's annual revolution through the zodiac, with its harmonious threefold loop rhythm, is illustrated in Figure 102. The planet's changing distance from the Earth is taken into account in the portrayal of the paths of the loops. The Earth is represented at the centre of the drawing; the directions of the twelve constellations of the zodiac are indicated on the outer circle. Mercury circles around the Earth in the course of a year, whereby its distance from the latter constantly changes; it approaches nearest to the Earth during its loops. As we

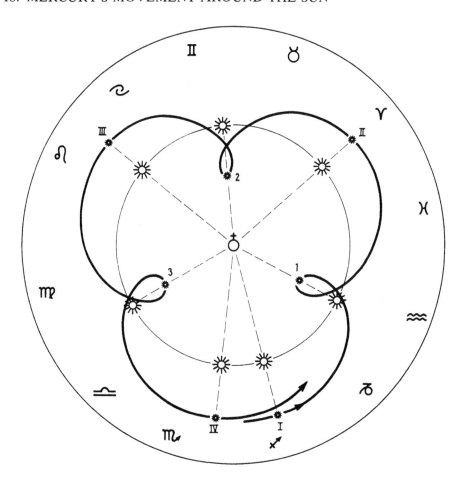

Figure 102. The geocentric orbit of Mercury in the course of one year (1961), comprising three synodic periods.

	Superior conjunctions		Inferior conjunctions
I	1961 Jan 5	1	1961 Feb 21
II	1961 May 2	2	1961 June 27
III	1961 Aug 14	3	1961 Oct 22
IV	1961 Dec 16		

have already seen, the positions of perigee and apogee coincide with the inferior and superior conjunctions, respectively. The positions of the Sun are drawn for each conjunction. It is obvious that Mercury stands between the Sun and the Earth during loop formation, at the inferior conjunctions, and that it is more distant than the Sun at the superior conjunctions. Mercury thereby crosses over the Sun's orbit in both directions, towards and away from the Earth, three times each year. Figure 99 illustrates this movement for a seven-year Mercury period (1961 – 67).

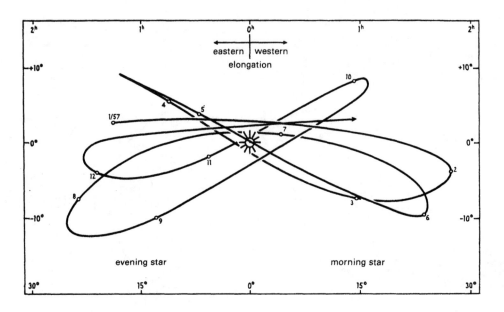

Figure 103. Mercury relative to the Sun in 1957.

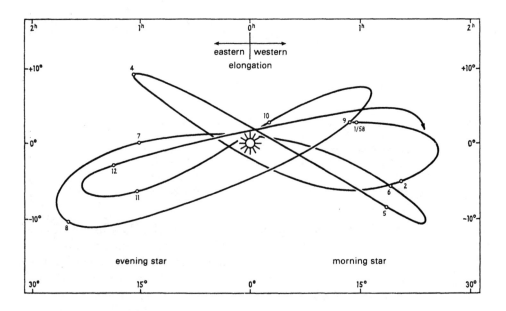

Figure 104. Mercury relative to the Sun in 1958.

18. MERCURY'S MOVEMENT AROUND THE SUN

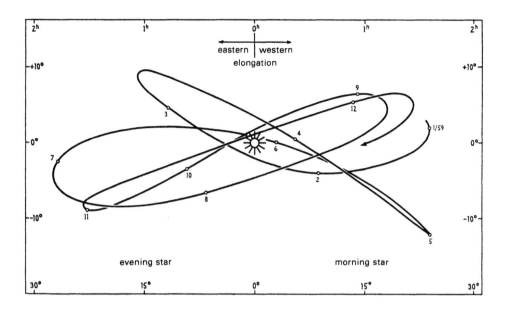

Figure 105. Mercury relative to the Sun in 1959.

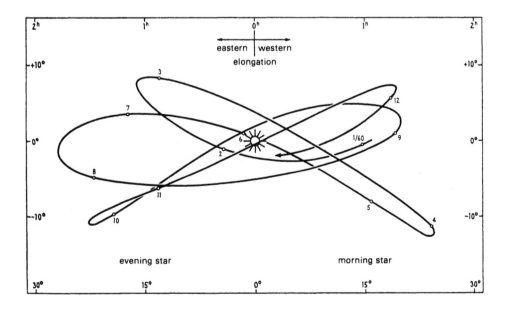

Figure 106. Mercury relative to the Sun in 1960.

143

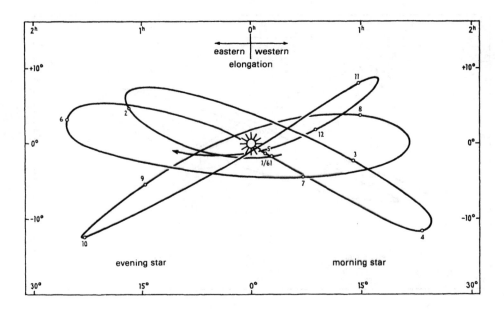

Figure 107. Mercury relative to the Sun in 1961.

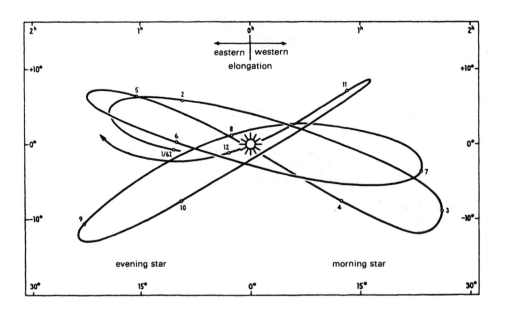

Figure 108. Mercury relative to the Sun in 1962.

144

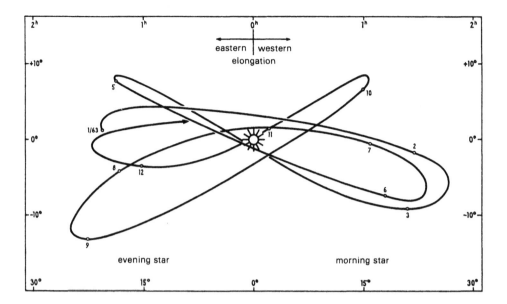

Figure 109. Mercury relative to the Sun in 1963.

As we have shown in the case of Venus, Mercury's revolution around the Sun can also be portrayed in another way by plotting the daily differences in ecliptical longitude and latitude, and connecting them in a curve. The Sun is taken as a fixed point. Figures 103 to 109 give the Mercury diagrams for seven consecutive years (1957–1963).

The planet's close circling about the Sun with the rapid alternation between evening and morning star is immediately apparent from the curve. Three times each year it is to be found on the eastern side of the Sun, and three times on the western side. The most distant points from the Sun in each loop represent the positions of greatest elongation. The variability of these distances can easily be seen, for instance 1958 (Figure 104), upper left and lower left.

In general, the greater an elongation, the greater is the likelihood of good visibility for Mercury. The conditions of visibility, however, do not depent on the extent of elongation alone, but also on the position of the two luminaries in the zodiac. If Mercury's declination is lower than that of the Sun, it will normally remain invisible. The higher the planet's declination relative to that of the Sun, the more favourable do the conditions of visibility become. The optimal conditions occur when a relatively great elongation coincides with a high declination, as for example in July 25, 1991 in the evening sky, or in January 1997 in the morning sky.

As for all cosmic events, a rhythmic variation can also be found to underlie the annual transformations of these pictures. If the Mercury diagrams are studied

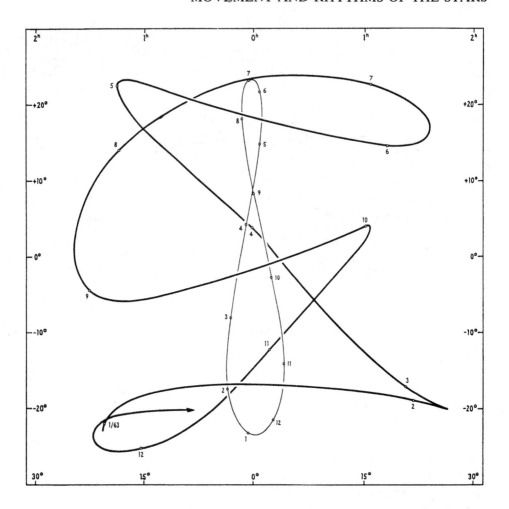

Figure 110. Mercury relative to the Sun in 1963, but also showing the annual movement of the Sun (compare Figure 109).

over many years, it becomes apparent that every seven years much the same curve recurs; in between lie transitional forms. One such cycle of transformations is given in Figures 103 to 109. The loops change both in height and width in a cyclic pattern, the following and preceding loop on the other side of the Sun always being at the opposite part of the cycle.

Figure 110 shows the Mercury diagram for 1963 as it appears when the changing daily positions of the Sun are taken into account. It has also been attempted to allow the spatial quality of the movement to come to expression.

19. Transits of Venus and Mercury

In exceptional cases Venus and Mercury can, during an inferior conjunction, pass across the Sun's disc, visible in the telescope as small black spots. Such transits, as they are called, are analagous to solar eclipses, which result when the Moon passes directly in front of the Sun.

Venus

If the orbital planes of Venus and Mercury exactly coincided with the Sun's (the ecliptic), such a transit would invariably take place at each inferior conjunction, where at the superior conjunctions the planet would always be occulted by the Sun. Neither, however, is the case, as the orbits of the two planets are inclined to the ecliptic. The orbit of Venus is inclined at $3\frac{1}{2}°$ to the latter, the crossing-points (nodes) being in opposite regions of the zodiac. Seen from the Earth the northern node (☊) lies in the Scorpion at 256° geocentric longitude (λ) while the southern node (☋) lies in the Bull at 76°. The Sun passes these positions on June 7 and December 9, respectively. Only when an inferior conjunction takes place in the vicinity of one of these nodes is it possible for Venus to pass directly over the Sun's disc. The maximum elongation is $1\frac{3}{4}°$.

As the planet's motion is always retrograde during an inferior conjunction, it passes into the Sun's disc on the eastern edge and leaves it on the western edge. The duration of the transit naturally depends on the distance of the planet's orbit from the mid-point of the Sun's disc. A so-called grazing transit will always be of brief duration; a central transit, by contrast, can last up to $7\frac{1}{4}$ hours.

Like the solar eclipses, the Venus transits cannot be observed everywhere on the Earth. Moreover, the duration of the passage varies with the geographical latitude of the observer.

Because the positions of the Venus conjunctions slowly advance in the zodiac in conformity with the period of 243 years, it follows that over a longer interval of time there are years when an inferior conjunction takes place near a Venus node. Eight years later the mutual positions have changed so little that the transit is repeated. During the progression of the positions of conjunction (see page 124) the ascending and descending Venus nodes are alternately passed. Two Venus transits therefore take place eight years apart at the ascending node, and then, after an interval of over a century, are followed by two similar transits at the descending node. The intervals between the individual transits are: 8 years, $121\frac{1}{2}$ years, 8 years and $105\frac{1}{2}$ years, after which the transits repeat themselves at the same intervals. (The nodes themselves wander so slowly through the zodiac that their movement is negligible).

The underlying rhythm is therefore 243 years and 2 days. The four transits occurring in this time could each be considered as a member of four series, as shown in the table below. In the twentieth century no transit of Venus takes place.

Transits of Venus

Interval	At ascending node		At descending node	
	1631 Dec 8	1639 Dec 4	1761 June 6	1769 June 3
243 years				
	1874 Dec 8	1882 Dec 6	2004 June 7	2012 June 5
243 years				
	2117 Dec 11	2125 Dec 8	2247 June 9	2255 June 7

Mercury

Mercury's orbit is inclined at 7° to the ecliptic. (With the exception of Pluto with 17°, this is the strongest orbital inclination of all the planets.) Geocentrically, the northern node lies in the Scales at 228° ecliptical longitude, and the southern node towards the end of the Ram at 48°. The Sun passes these parts of the zodiac on November 11 and May 9, respectively. If, therefore, an inferior conjunction of Mercury takes place in the immediate vicinity of one of these two dates, a Mercury transit takes place. The planet thereby passes across the Sun's disc as a small black point, which can be observed telescopically. The transits in November take place at the northern node; those in May at the southern node. In order for a transit to occur the planet may not have more than $3\frac{1}{2}$° elongation from the node.

Like Venus, Mercury is always retrograde during an inferior conjunction, so that the planet makes its first contact with the eastern edge of the Sun, then moving across to the west. As a result of Mercury's more rapid movement, its transits are generally of briefer duration than those of Venus. The maximum length of a central transit is about $5\frac{1}{2}$ hours (Figures 111 and 112). Because Mercury is closer to the Sun than Venus, the conditions of observation vary only slightly for different localities of the Earth.

In a period of 217 years, after which the Mercury transits are repeated fairly exactly, roughly twenty transits take place in November, and roughly nine in May. The transits in November are more frequent than those in May because Mercury is nearer the Sun at this time of year during its inferior conjunctions, owing to the high eccentricity of its orbit. The intervals for the November transits are: seldom 6, mostly 7 or 13 years; for the May transits 13, very seldom 20, otherwise 33 years. A November transit always follows $3\frac{1}{2}$ years after a May transit.

148

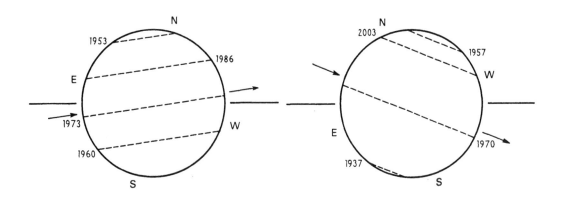

Figure 111. Transits of Mercury in November. The horizontal line is the ecliptic.

Figure 112. Transits of Mercury in May. The horizontal line is the ecliptic.

Six transits usually take place in 46 years (and 13 or 14 in a century). They follow one another at the intervals of 7, $9\frac{1}{2}$, $3\frac{1}{2}$, $9\frac{1}{2}$, $3\frac{1}{2}$ and 13 years, thereafter repeating in similar sequence. Now and again there is an interruption in the cyclic sequence, as for example by the transit of 1957.

Mercury's planetary period of 46 years (see Chapter 17) can also be considered as a unified rhythm for the Mercury transits, comparable to the Saros cycle for Sun and Moon eclipses. The transits can be arranged into six series, whose individual members return with only slight variations after 46 years. Only two of these cycles take place at the descending node; the other four occur at the ascending node.

The 46-year periods do not go on for ever, but have, like the Saros periods, a certain 'life-span'. For the transits in May there are generally ten members lasting 414 years; for instance, May 1523 to May 1937, or May 1740 to May 2154. A November series embraces more than double this time (about twenty members, lasting 874 years).

The twentieth century forms an exception within these great cycles, insofar as two cycles come to an end (May 1937 and November 1999) and two new ones begin (May 1957 and November 1993).

The May transits at the descending node migrate in the 46-year period from north to south on the Sun's disc. Those in November at the ascending node wander from south to north over the Sun.

The period of 217 years is not without effect on the 46-year period and the individual series. For example, a May or November cycle begins and ends 217 years before and after another such cycle begins or ends.

Forty-six-year periods of Mercury transits

Nov	May	Nov	May	Nov	Nov	May	Nov	May
							1605	1615
1618	1628	1631			1644		1651	1661
1664	1674	1677			1690		1697	1707
1710		1723			1736	1740	1743	1753
1756		1769		1776	1782	1786	1789	1799
1802		1815		1822		1832	1835	1845
1848		1861		1868		1878	1881	1891
1894		1907		1914		1924	1927	1937
1940	Nov	1953	1957	1960		1970	1973	
1986	1993	1999	2003	2006		2016	2019	
2032	2039		2049	2052		2062	2065	
2078	2085		2095	2098		2108	2111	

Johannes Kepler was the first to predict such planetary transits, on the basis of his own planetary tables. In 1629 he forecast two transits for the year 1631: one of Mercury on November 7, and one of Venus on December 6. As it is most exceptional for two such transits to occur in the same year, let alone only a month apart, Kepler attributed a special significance to this event, and addressed himself to all astronomers and friends of the stars in an appeal. Kepler himself was unable to observe these rare phenomena, as he died in November 1630. Nonetheless, on the day of Kepler's prediction various observers actually succeeded in seeing a transit of Mercury for the first time.

In the year 1769 this double event recurred with Venus in June and Mercury in November. Since that time such an event has not taken place again, and will not do so until the beginning of the twenty-third century.

20. The loop-formation of Venus and Mercury

Figure 113 illustrates the loop orbits of the planets Venus and Mercury in their separate periods of 584 and 116 days, respectively. Points I and II define the positions of consecutive superior conjunctions to the Sun. In the middle of each loop falls an inferior conjunction. The large overlap of the Venus movement appears to be a kind of compensation for, or complement to the broad portion of the sky which Mercury does not cover in its corresponding movement.

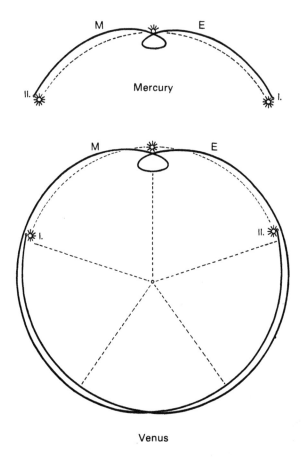

Figure 113. The loops of Mercury and Venus in their synodic periods. I and II are successive superior conjunctions. Morning visibility (M) and evening visibility (E) are indicated.

Venus

In the course of her eight-year cycle Venus describes five loops in the zodiac at the times of the inferior conjunctions. In their eight-yearly recurrence, the five loops assume very nearly the same form and position. No other planet retains the forms of its orbit as consistently as Venus. The loop-pentagram gradually moves in the zodiac (see page 124) as though it were slowly rotating, so that the loop in the Lion, for example, is slowly wandering in the direction of the Crab (Figure 83). The fact that the various stations of Venus' movement return $2\frac{1}{3}$ days earlier after eight years, corresponds to a clockwise displacement of the loops by about $2°3$.

Figure 114 shows the gradual movement of a Venus loop, and the accompanying change in form, when it returns at eight-year intervals to the same

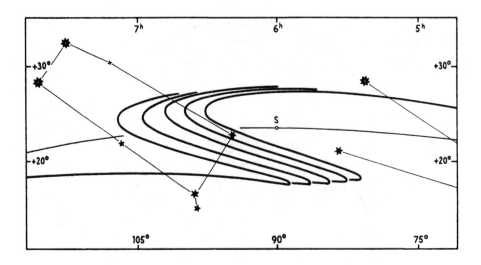

Figure 114. The displacement and transformation of a Venus loop at eight-year intervals in the same constellation (Twins). Summer solstice (S) on the ecliptic is indicated.

region of the zodiac. The continual wandering of all the loops gradually covers all portions of the zodiac with an intricate network of loop-patterns. This is already the case after 243 years, during which time the loop or conjunction-pentagram rotates by a fifth revolution, or 72° (see Chapter 15). In this time all the transitional forms of the loops appear, although strongly overlapping. The metamorphosis follows in principle the same laws for both planets, the difference lying in the specific position of the nodes. The open, or zigzag loops take place for Venus in the constellations Bull and Scorpion. Venus' loops are larger in dimension than those of all other planets, with the exception of Mars, whose loops only occasionally exceed those of Venus. The average length of the retrograde movement is 17°.

Mercury

As we have seen in Chapter 17, Mercury describes three loops each year; those of the following year lie close to those of the past year, but without overlapping. In this manner, Mercury distributes its loop patterns over the whole zodiac in only a few years. Such a series of twenty-two loops described in as many synodic periods, is completed in seven years. Once again, the same principle of transformation underlies these phenomena. The metamorphosis of the Mercury movement can be seen in Figure 98 for a seven-year cycle (1961 to 1967).

Especially striking are the rapidity and the manifold quality of Mercury's movements, which nonetheless follow a regular pattern. It can be seen from Figure 98 how the individual loops show the various typical forms corresponding directly to their different positions in the zodiac. Beginning with a nearly symmetrical loop

152

in the Crab, there follow asymmetrical loops to either side, which are nearly mirror images of each other. These forms at first become narrower and more pointed, and then pass over into open zigzag forms about 90° from the starting-point. The loops then once more become narrow, before taking on closed, rounded forms, and returning at last to symmetry in the opposite portion of the zodiac, in the Goat.

In their totality the loops display a definite orientation in the zodiac. The retrograde portions of the orbits cross over the ecliptic in the *S* or *Z*-forms. In the Crab, by contrast, the loops assume their greatest southern latitude from the ecliptic (in Figure 98 the loops are turned inwards), while in the Goat they rise to the north of the Sun's orbit (outwards). The whole ring of loops thus takes on a distinct tilt to the ecliptic, with two crossing-points (nodes), marked by the *S* and *Z*-formed loops in the constellations of the Ram and the Scales.

The origin and the ordered sequence of the different loop-forms is a result of the changing position of Mercury's orbit. On the right-hand side of Figure 115 are shown from *a* to *h* eight different positions of Mercury's orbit around the Sun, as it appears from the Earth. The uppermost position *a* is attained each year on August 10. If Mercury's positions relative to the Sun were plotted for this date in many different years, they would all gather along an elliptical ring around the Sun. The lower half of the ellipse represents the portion of Mercury's orbit which is nearer the Earth; the upper half is turned away from the Earth, behind the Sun. Correspondingly, to each day of the year belongs a different situation of Mercury's orbit. Position *c* (November 11) represents the moment when the view from the Earth falls directly along the edge of the momentary orbital ring. From this position (positions *d, e, f*) onwards, the portions of the orbit nearer the Earth lie above the Sun; in other words, our view falls on the underside of the orbital plane, whereas in *a* and *b* the orbit is seen from above. After attaining its most open position, *e* (February 7), the ring once more becomes narrower, until at *g* (May 9) it appears again as seen from the edge, but with opposite tilt to the position six months earlier.

In connection with the changing positions of Mercury's orbit, which may be deduced directly from the celestial phenomena, the corresponding form-types of the planetary loops are illustrated on the left-hand side of Figure 115.

If we assume the orbit position *a* to be fixed, Mercury would during the course of a synodic revolution, pass successively through points 1 to 12. At the same time, the Sun changes its position in the zodiac. Mercury's circling movement around the Sun is therefore modified from our point of view by a lateral progression along the ecliptic. As Mercury passes through the basic form *a*, the points 1–12 move at the same time horizontally from right to left. Out of these two movements the loop-curve 1 results. In the same manner all the other loops may be deduced from the simultaneous lateral movement of the adjacent orbits. The drawings do not take into account the fact that the positions of the orbits are

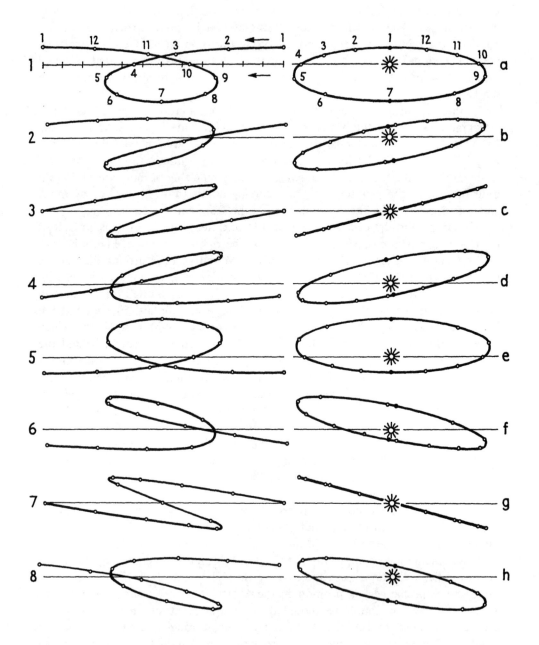

Figure 115. The position of Mercury's orbit around the Sun at different times of year, and the corresponding loop forms.

a Aug 10   c Nov 11   e Feb 7   g May 9
b Sep 26   d Dec 25   f March 26   h June 26

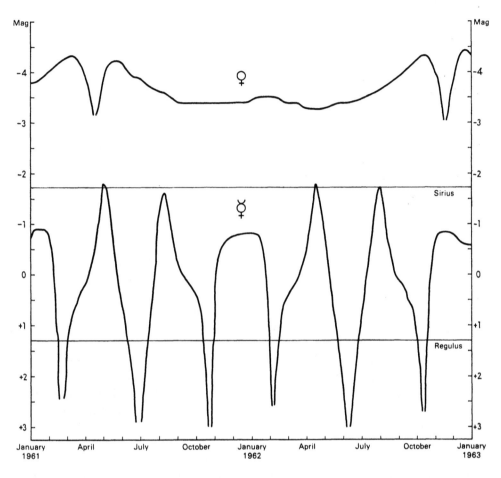

Figure 116. Graph of variations of brightness of Venus and Mercury. For comparison, the magnitude of Sirius and Regulus are shown.

continually changing; only the individual stages are represented. They can, however, serve as a practical key to understanding how any given Mercury loop arises. As an example, we may take the loop of March/April 1966 (Figure 98, in the Fishes), and the loop-form which results from the position of Mercury's orbit in March (Figure 115, 6 and *f*) with the help of our construction.

Each year Mercury singles out three loops from its whole loop-organism and traces them out in the sky, though all are potentially present. Irrespective of the typical forms, individual Mercury loops display slight deviations and irregularities, particularly with respect to their length, as we mentioned earlier on page 134.

155

Comparisons

Venus, like Mercury, circles around the Sun. Its orbit, however, is more extended, and the corresponding rhythms, therefore, longer. The way in which its loop-forms arise can be derived in the same manner as for Mercury.

From a certain point of view, Venus, Mercury and the Moon can be thought of as belonging to a single group.

All three luminaries can stand between the Earth and the Sun; they all possess the ability, to a greater or lesser extent, to eclipse the Sun's disc.

The most striking singularity of the Moon, the change of phase, can be observed telescopically for Venus and Mercury. This transformation is of longest duration for Venus (1 year 7 months, or 584 days); for the Moon it is shortest (29½ days); between these extremes stands Mercury (116 days).

This relationship is also reflected in their brightness. Their magnitude is plotted on a graph (Figure 116). Beginning with a superior conjunction, Venus undergoes a gradual increase in brightness until it reaches its maximum intensity as an evening star; her light then diminishes rapidly and declines at last into complete darkness at the inferior conjunction. (In extreme cases a very fine crescent can be observed along the planet's northern or southern edge. Thereupon follow the rapid ascent to maximum brightness in the morning sky, and the slow fading as the planet approaches its superior conjunction with the Sun. Here, however, it does not lose its luminosity altogether, as at the inferior conjunction. This whole process is extended over the synodic period of 1 year and 7 months. Mercury undergoes the same transformation three times each year. The Moon shows in the course of 29½ days a very steep and strong ascent of its light-intensity to Full Moon, followed by a steep descent which is interrupted at New Moon. The scale of the graph, however, is too small to display these fluctuations, for the Full Moon attains a magnitude of −12.5.

The three luminaries display, as it were, a 'lunar character', which is, of course, most strongly manifested by the Moon itself; Venus inclines rather to the nature of the outer planets, with which we shall be concerned in the following chapters. She is accordingly the most tranquil and steadfast of the three, while the Moon is the most rapidly moving and the most difficult to apprehend. Mercury assumes a middle position.

21. Some phenomena common to Saturn, Jupiter and Mars

The star-filled heavens seem all the more impressive when the mild, pale glimmer of Saturn, the radiant, yellow-white light of Jupiter, or the deep orange-reddish glow of Mars are visible and stand out prominently among the twinkling hosts of brighter and paler fixed stars. Among these three luminaries, Jupiter is the most sovereign: wherever and whenever it appears in the sky it outshines all other stars in its vicinity. Saturn, by contrast, holds its light within more modest bounds, although it too can exceed nearly all the fixed stars in brightness.

Only from time to time does the reddish lustre of Mars become dominant in the night sky, namely during the periods of its best visibility, which recur at two-year intervals.

With respect to their phenomena and conditions of visibility, Saturn, Jupiter and Mars display a number of marked contrasts to the inner planets Venus and Mercury. The two latter planets are limited in their movements to a definite region to either side of the Sun, whose boundaries are fixed by the points of greatest elongation. Saturn, Jupiter and Mars, by contrast, can assume any lateral elongation to the Sun, for instance quadrature (90° separation), and opposition (180° separation, when the planet and the Sun stand in opposite parts of the zodiac).

These relationships make it possible for the three outer planets to be visible during the whole night, whereas the visibility of Mercury and Venus is limited to the morning or evening hours.

Each period of visibility of these three planets proceeds regularly in the following sequence. At the time of a *conjunction* with the Sun, Saturn, Jupiter and Mars remain invisible. The planet, in this position, accompanies the Sun during the course of the day, rising and setting together with the latter. After a number of weeks, the planet becomes visible as a dimly shining point of light in the eastern morning sky (heliacal rising). From this time on the planet rises earlier and earlier before the Sun, until it gradually dominates the second half of the night. At the same time it increases in brightness. The exact position and height of the arc along which it ascends each night varies, depending on the constellation of the zodiac in which it is located. After some time, the planet rises already before midnight, and finally it can be seen near the eastern horizon at sunset; it is then standing in *opposition* to the Sun.

The planet has now attained its best visibility and greatest brightness. In the first hours of the night it ascends in the eastern sky, reaches its culmination at midnight, and descends during the second half of the night towards the western horizon. At the time of opposition the planet shines throughout the night, from sunset to sunrise. In the following weeks and months the period of visibility becomes shorter again. It becomes increasingly limited to the first half of the night, and finally to the evening hours shortly after sunset. Now the planet is to be found over the horizon in the south-west, west or north-west, depending on its position in the zodiac. The whole period of visibility, which began with the heliacal rising, ends with the *heliacal setting*, or the disappearance of the planet in the evening twilight.

This cycle of phenomena, or synodic period, can be characterized for Saturn, Jupiter and Mars in the following stages:

Planet's position relative to the Sun	*Planet's visibility*
Conjunction	Invisible
Heliacal rising (west of Sun)	Appears in morning
Western quadrature	Second half of night
Opposition	Throughout night
Eastern quadrature	First half of night
Heliacal setting (east of Sun)	Disappears in evening
Conjunction	Invisible

The rhythmic recurrence of the same positions, that is the synodic period, takes place in the case of Saturn and Jupiter in slightly more than a year, whereas Mars requires double the time: more than two years.

As Saturn, Jupiter and Mars are visible for months at a time in the night sky, their movements in the zodiac can easily be followed against the background of the fixed stars. If the planets' positions in the sky are followed systematically for several weeks it can be seen how they progress past the neighbouring stars. In contrast to Mercury and Venus, which remain in the broader vicinity of the Sun, alternately moving ahead of and falling behind its annual movement through the zodiac, Saturn, Jupiter and Mars display quite different relationships.

They move considerably more slowly through the zodiac than the Sun. Whereas

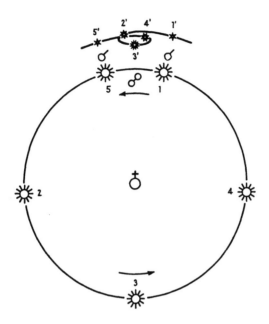

Figure 117. The movement of Sun and Jupiter in the zodiac during one synodic period of the planet.

the latter progresses by about 30° each month along the ecliptic, Jupiter requires a year to travel the same distance, and the even slower Saturn two and a half years. Mars' speed, however, is closer to that of the Sun.

The characteristic differences in speed have the result that the faster-moving Sun periodically overtakes the three slower planets and, in the course of its annual revolution, continually changes its angle to the planets. This results in ever-changing relative positions of the Sun and each of the three planets, in which the aspects of conjunction, quadrature and opposition follow each other in regular succession.

Seen over longer periods, Saturn, Jupiter and Mars move forward through the zodiac, that is, in the same direction as the Sun in its annual movement. But this *direct* movement of the planets is periodically interrupted, and for a time their visible motion against the heavens becomes *retrograde*. This alternating forwards and backwards movement gives rise to the formation of loops, which may be open or closed, like those of Venus and Mercury. Each planet describes one such loop during a synodic period, the retrograde portion of which is always completed around the time of opposition to the Sun. In the middle of the direct portion of the orbit, between two loops, lies the point of conjunction.

159

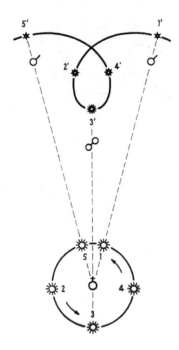

Figure 118. The changing distance between Earth, Sun and Jupiter during the latter's synodic revolution.

The interplay of the simultaneous movements of the Sun and the planet in the zodiac during one synodic period is illustrated for Jupiter in Figure 117. The Sun, proceeding from one conjunction with the planet (1 – 1') advances past positions 2, 3 and 4, until, after a year, it has circuited the whole ecliptic. The planet, in the meantime, has moved over the corresponding positions 2', 3' and 4'. The Sun must now overtake the planet, so that the next conjunction takes place at the position 5 – 5'. The position 2 – 2' represents the western or first quadrature, 3 – 3' the opposition, and 4 – 4' the eastern or second quadrature between Sun and Jupiter.

This illustration makes clear the harmony of movements which take place between Sun and planet during each synodic period. At the time of conjunction both move in a common direction, but at different speeds. As the Sun then moves farther away from the planet and passes on to the opposite part of the zodiac, the planet accompanies, as it were, the Sun's circuit with a retrograde movement. From this aspect the loop movement appears as a kind of resonance of the planet to the Sun's annual course through the zodiac. This resonance becomes still more pronounced when we consider the planet's varying distance from the Earth.

At the time of conjunction Saturn, Jupiter and Mars are beyond the Sun. At

160

this time they stand farthest from the Earth, at apogee. Thereafter their distance from the Earth decreases, until they reach their perigee at opposition. Figure 118 illustrates these relationships for Jupiter. When the Sun is in opposition, Saturn, Jupiter and Mars are closest to the Earth; in describing their loops, they show at the same time an orientation to the annual movement of the Sun.

The lesser distance of these three planets from the Earth at the time of their opposition also results in their greatest brilliancy coinciding with their longest visibility.

22. The rhythms of Saturn and Jupiter

If the phenomena and movements of Saturn and Jupiter are followed in the sky, it will be noticed that they change their position fairly slowly. During their period of visibility, Saturn and Jupiter shine from one constellation for many months. Jupiter takes a whole year to move through a single constellation; its complete course through the zodiac therefore requires almost twelve years. Saturn, still slower, takes almost thirty years to complete its course through the constellations.

Compared with the Sun, Saturn and Jupiter move exceptionally slowly. As the Sun moves through the ecliptic once each year, its visible distance in the sky (the elongation) from Saturn or Jupiter changes rather rapidly. From this movement ensue the changing mutual positions which we have described above with their most prominent stages: conjunction, quadrature and opposition between Sun and planet. These positions are attained in regular sequence; at rhythmic intervals the Sun always returns to the same position relative to the planet.

The synodic period
For *Jupiter*, the interval from one conjunction to the next, the synodic period, is roughly 1 year and 1 month, or more exactly, 399 days. For *Saturn* the corresponding time is slightly less, as the Sun catches up on the slower planet more quickly. Saturn's synodic period is one year and about two weeks, or more accurately, 378 days.

As we have already seen, during the synodic period of a planet all possible positions relative to the Sun occur, giving the changing length of visibility. At the time of conjunction with the Sun the planet is not visible. As its distance from the Sun increases, it becomes visible at first for a shorter, then for a longer time, until the optimal visibility is attained at opposition.

The cycle of phenomena and the periodic alternation of visibility and non-visibility occur in the following sequence for Saturn and Jupiter:

Planet's position relative to the Sun	Period (days) Saturn	Jupiter	Planet's visibility
Conjunction			Invisible
	98	114	
Western quadrature			Second half of night
	91	85.5	
Opposition			Throughout night
	91	85.5	
Eastern quadrature			First half of night
	98	114	
Conjunction			Invisible
One synodic period	378	399	
	1y 13d	1y 34d	

The difference between the synodic revolution and the solar year (13 or 33 days) causes a shifting, such that the same phenomena of Saturn and Jupiter occur at later dates.

The same conditions of observation for Saturn occur about half a month later each year. This displacement causes the period of optimal visibility to wander slowly through the seasons: from spring to summer, and into autumn and winter, until it returns to its original position after roughly 30 years (the sidereal revolution).

Jupiter shows a corresponding annual delay of 33 days in the recurrence of its positions. The progression of its period of optimal visibility through all the seasons takes place in about 12 years.

Tables 6.7 and 6.8 show the phenomena and periods of visibility for Jupiter and Saturn from 1981 to 2010.

The sidereal period

If the movement of Saturn or Jupiter against the fixed star background is examined more closely a rhythmic alternation will be found between direct and retrograde motion, in other words, a loop-motion of the planet (see Figure 117). The place where the planet reverses its direction in the zodiac is called the *stationary point*. Saturn and Jupiter both reach these points three to four weeks after their western and before their eastern quadrature. From the first stationary point, which marks the beginning of the retrograde movement, to the centre of the loop, where the opposition with the Sun takes place, nearly seventy days pass for Saturn, and for Jupiter about sixty days. After an equal interval the second stationary point is attained. At this point of the loop the retrograde motion changes over again to direct motion.

162

Saturn requires roughly 140 days, or $4\frac{1}{2}$ months, for the completion of its retrograde motion, and Jupiter 120 days, or 4 months. The segment of retrograde motion is only 7° for Saturn; Jupiter's is on average 10°.

As the loops of Saturn and Jupiter always take place during optimal visibility, the planets' changing positions can easily be followed, and the whole loop-formation observed.

Jupiter's longer rhythms

From one conjunction with the Sun to the next, the positions of Saturn and Jupiter advance continually in the zodiac, which results in the loops occurring in different parts of the latter. Figure 119 shows this progression for Jupiter for one sidereal revolution (1959–1970). The planet moves along by roughly 33°, or one eleventh of the zodiac annually. A complete revolution contains eleven loops, which are completed in eleven synodic periods, or about twelve years. More precisely, 11 synodic periods of Jupiter are 4388 days, whereas 12 years are 4383 days. The discrepancy is therefore only 5 days.

This difference is sufficient, however, to bring about a further metamorphosis. At the beginning of each new twelve-year period Jupiter's loops have progressed by about 5° to the east, bringing about a delay of 5 days. After 83 years Jupiter has described 76 loops, returning to a position only about 8′ east of its starting point. This corresponds to a delay of about 3 hours.

Figure 120 shows Jupiter's changing distances from the Earth during one revolution through the zodiac. The exact distances are given in Table 4.1.

The twelve-year period of Jupiter's progression through the zodiac may be thought of as a 'Jovian year', by analogy to the solar year. During this period the arcs of Jupiter's daily motion rise and sink, and its directions of rising and setting change. During a solar year roughly the same metamorphosis takes place in Jupiter's diurnal arcs, as for those of the Sun in a month. Over a period of six years, when Jupiter is moving through the upper portion of the zodiac, it ascends to relatively high culminations, as for example, from 1987 to 1992, and again from 1999 to 2004. In addition there is a period of especially good visibility during the oppositions, at which time Jupiter, owing to its high position in the zodiac, is above the horizon for more than twelve hours. During the following six years we find Jupiter in the lower constellations of the zodiac. It then remains close to the horizon in the southern sky. Its visibility is less favourable and of shorter duration.

The rising and sinking of the arcs of daily movement is not uniform, but, as with the Sun, follows a harmonious oscillation (see Chapter 6) which in this case is extended over twelve years. Exact observation, however, shows that the ascent and descent is not uniform, but finely differentiated in a rhythmic pattern arising from the planet's looping motion. During Jupiter's progression from the Archer to the Twins, the diurnal arcs expand so long as the planet's motion is direct, and contract again slightly when it becomes retrograde. Then there comes a new

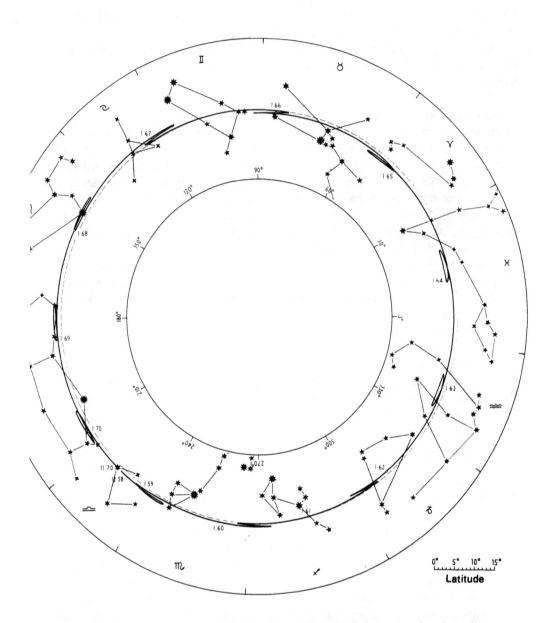

Figure 119. The loops of Jupiter 1959–70. The small circles indicate the planet's position at the beginning of each year. The height of the loops is slightly exaggerated. The loops in the next twelve-year period (1971–82) take place about 5° to the east of those shown here. The loop for 1971, for example, is slightly to the right of the loop for 1959, overlapping it by half its length. Ecliptic: ————.

164

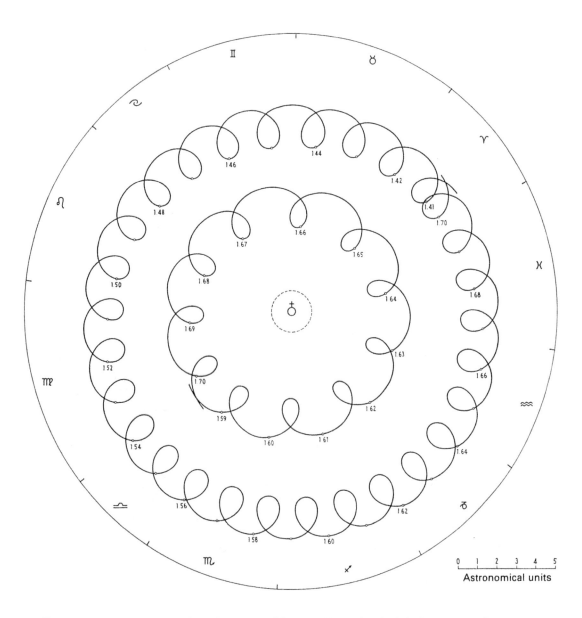

Figure 120. The geocentric orbits of Saturn and Jupiter (1940–70 and 1959–70 corresponding to Figures 122 and 119, respectively). The directions of the constellations are shown on the outer edge. The dashed circle indicates the Sun's orbit.

165

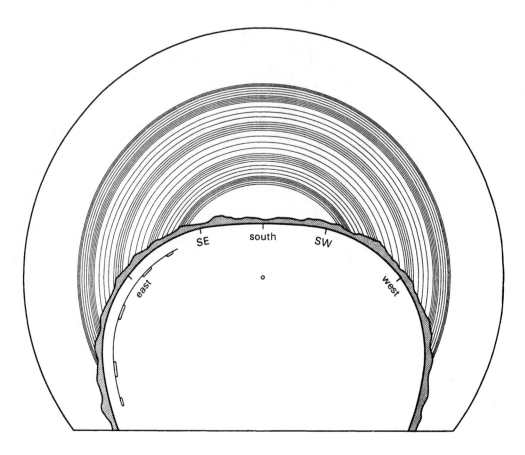

Figure 121. The rhythmic expansion and contraction of the diurnal arcs of Jupiter in its passage from the Archer to the Twins.

expansion, followed by a slight contraction, and so on. The six-year ascent takes place, as it were, stepwise, with annual rhythmic contractions. This expanding and contracting movement is reflected in the planet's directions of rising and setting (Figure 121). When Jupiter again descends from the Twins to the Archer the situation is exactly reversed: during the direct motion a general contraction takes place, followed by a slight expansion during the retrograde motion. As the daily arcs pass over continuously into one another, a spiralling rising and sinking movement results.

Saturn's longer rhythms

Saturn, in accordance with its slower movement, remains in one half of the zodiac for roughly fourteen to fifteen years. From 1980 to 1996 it wanders through the southern constellations from the Virgin to the Fishes; from 1986 to 1990 it is

166

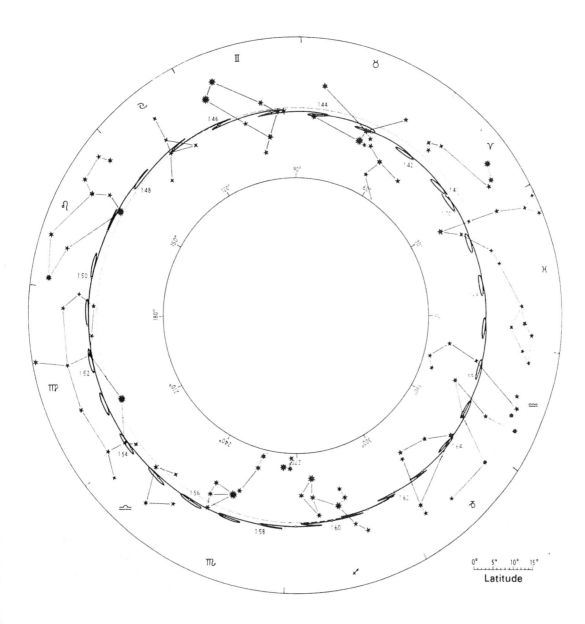

Figure 122. The loops of Saturn 1940–70. The small circles indicate the planet's position at the beginning of each year. The height of the loops is slightly exaggerated. For the period 1971–2000 the loops fall roughly between the loops shown here. The loop for 1971, for instance, lies between the loops of 1941 and 1942.

167

especially low in the sky in the northern hemisphere, while it culminates high in the northern sky of the southern hemisphere. From 1996 to 2010 it wanders through the northern constellations from the Fishes to the Virgin; from 2003 to 2005 it is especially high in the southern sky (in the northern hemisphere). Because of its slow progression in the zodiac, Saturn remains for an average of two and a half years in the same constellation, where it describes two or three consecutive loops. Figure 122 illustrates the loops of one sidereal revolution of Saturn (1941–1970). In this period of 29 years Saturn completes 28 synodic periods, describing 28 loops. Figure 120 shows the changing distances between Saturn and the Earth during the same period (for the numerical values, see Table 4.1 in the Appendix).

As the period of 28 synodic Saturn periods is slightly shorter than 29 years (28 synodic Saturn periods = 10 586½ days, 29 years = 10 592 days), the new 29-year period begins 5½ days earlier in the seasons than the preceding period. The loops are correspondingly shifted by $5°3$ to the west of those at the beginning of the previous period. In Figure 122 this change can be distinguished if we compare the two Saturn loops of 1941 and 1970. After roughly double this time, 59 years with 57 synodic periods, a fairly exact equalization takes place. The same Saturn configurations are then repeated, only 2 days later in the year. Saturn's course through the zodiac, or one sidereal revolution, takes 29 years, 167 days, or 10 759.2 days (see Table 4.7 of the Appendix).

Like Sun and Moon, the planets move at uneven speeds through the zodiac. These variations are least noticeable with Jupiter and Venus; in the case of Mars, however, they are unusually prominent. Saturn, too, shows strong fluctuations. Its movement through the upper part of the zodiac, from 1996 to 2010, takes place in roughly 14 years, less than half of the sidereal revolution, whereas its journey through the lower portion of the zodiac, from 1980 to 1996, over a period of 16 years. Correspondingly, 15½ loops fall in the lower half of the zodiac, while the remaining 13½ are distributed with a somewhat wider separation over the upper half.

As with Jupiter, the gradual 15-year expansion of Saturn's diurnal arcs is interrupted during the retrograde motion by a slight contraction. In a relatively short time Saturn covers a small strip of the sky three times: in expansion, contraction, and then expansion again. In this manner, the spiralling fabric of Saturn's diurnal arcs is divided into 14 more diffuse and 14 denser bands. The breadth of these zones changes (also for Jupiter) according to the portion of the zodiac in which the retrograde motion occurs. The rising and setting points also change accordingly.

23. The rhythms of Mars

The synodic period

Of all the outer planets, Mars' succession of phenomena and rhythms of movement are most difficult to comprehend. On one hand, Mars has a synodic period of more than two years, and thus requires a greater length of time than any other planet to return to the same stages of its cycle. On the other hand, its motions, whether considered spatially or temporally, show irregularities and variations to a degree not found for any other planet.

So, for example, the time between two consecutive conjunctions or oppositions to the Sun, that is, the synodic revolution, can vary by almost 50 days. The period fluctuates between 2 years 34 days, and 2 years 80 days. The average value is 2 years 50 days (780 days).

Half a synodic revolution, marked for example, by the progression from a conjunction with the Sun to an opposition, therefore takes place in slightly more than a year. Parallel to this is the gradual transition from invisibility to optimal visibility. Mars' visibility increases for a whole year, and decreases the following year.

Before and after every conjunction with the Sun, Mars remains invisible in the latter's neighbourhood for a good half year (six to seven months). At the end of such a period, or about a quarter year after the conjunction, the heliacal rising takes place, the planet's first appearance in the morning sky. Mars appears at first as an insignificant object of low luminosity, similar to a star of the first magnitude; it is therefore not prominent. Not until half a year later, or about eight months after the conjunction, does Mars reach its western quadrature with the Sun, at which it is visible in the eastern sky from about midnight onwards. Over the next three to four months, until the opposition, the nightly period of visibility increases steadily. At the same time its reddish glow waxes. The brilliance of Mars can grow by more than four classes of magnitude, that is, to a maximum of about sixty-fold. At a bright opposition Mars outshines all the fixed stars, and of the planets only Venus attains a greater brilliance. To be sure, Mars only lights up for a relatively brief time to such an impressive radiance; a few weeks later it declines strongly. Finally, after a long period during which its visibility is restricted to the evening sky, it disappears in the west at its heliacal setting. This takes place about a quarter year before the next conjunction with the Sun.

The most important positions of Mars during its synodic period can be divided into the following stages:

Mars' position relative to the Sun	Average interval		Visibility
Conjunction			Invisible
	¼ year		
Heliacal rising		9½ months	Appears at dawn
	½ year		
Western quadrature			Second half of night
	¼ year	3½ months	
Opposition			Throughout night
	¼ year	3½ months	
Eastern quadrature			First half of night
	½ year		
Heliacal setting		9½ months	Disappears at dusk
	¼ year		
Conjunction			Invisible

The sidereal period

The change in the relative positions of Mars and the Sun during the two-year synodic period is governed by their different speeds through the zodiac. Figure

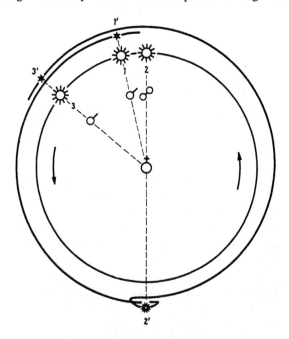

Figure 123. The movement of Sun and Mars in the zodiac during one synodic period of the planet.

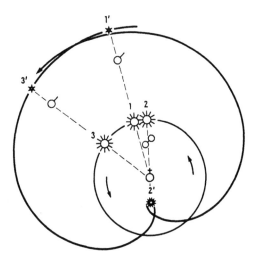

Figure 124. The changing distance between Earth, Sun and Mars during one synodic period.

123 gives a general view. Mars moves, on the average, only half as fast through the zodiac as the Sun. Beginning with a conjunction (1 – 1'), the Sun completes almost a whole revolution through the zodiac in the time which Mars requires to reach the opposite portion of the ecliptic: here the opposition to the Sun takes place (2 – 2'). Figure 124 shows the corresponding relation of distances from the Earth, which will be discussed further below.

As Mars' synodic period is, on the average, two months longer than two years, the meeting-places of the planet with the Sun shift in the zodiac. A similar displacement occurs for the positions and dates of the oppositions. A general survey is given for the years 1981 to 2010 in Table 6.6 in the Appendix.

Similarly, the loops occurring at the time of opposition slowly shift. Figure 125 shows the positions of the Mars loops for a fifteen-year period (1954–1969). The path of direct motion through the whole zodiac between the individual loops is not shown. As can be seen from the plate, the loops progress over rather irregular distances every two years. Further, in different parts of the zodiac the loops have greatly varying lengths. On the average, the retrograde portion of the path is 15°. The longest loops in the constellations of the Lion and the Virgin are more than $19\frac{1}{2}$° in length, while the smallest, around the Water-Bearer, are only $10\frac{1}{4}$°. The duration of the retrograde motion varies accordingly. On the average, the stationary points in the zodiac are reached 37 days before and after the opposition. The average length of the retrograde loop-movement is 74 days, or $2\frac{1}{2}$ months. This period, however, can vary between 60 and 82 days. This variation at the same time defines the period in which Mars' brightness and length of visibility are at their optimum.

171

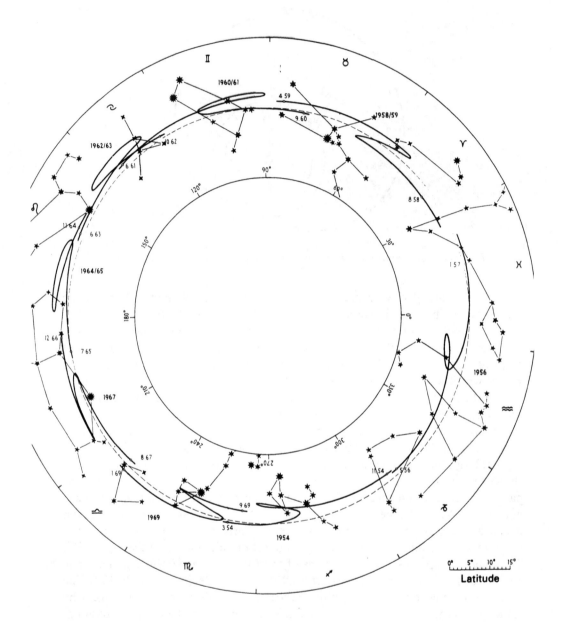

Figure 125. The loops of Mars (1954–69). Between two successive loops Mars wanders once through the whole zodiac.

172

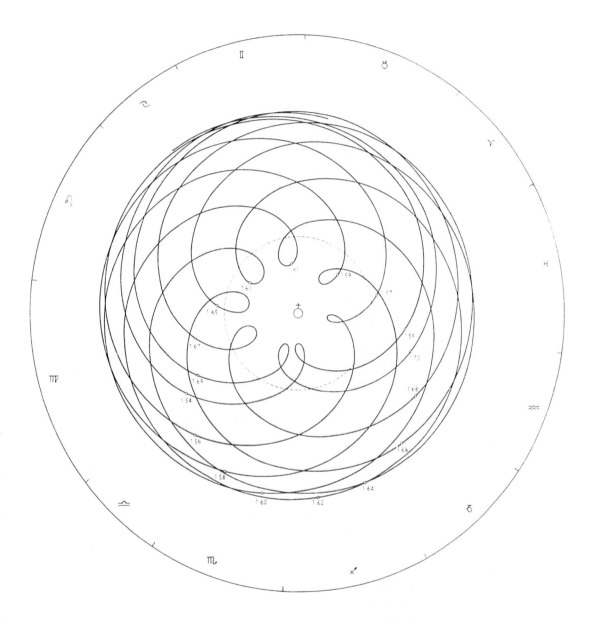

Figure 126. The geocentric orbit of Mars 1953–70, corresponding to Figure 125. The small circles show the planet's position at the beginning of each year. The directions of the constellations are shown on the outer edge. The dashed circle represents the Sun's orbit.

After seven completed loops, or seven synodic periods (about fifteen years) the Mars loops have progressed once through the whole zodiac. Thus every fifteen years a loop is repeated in the same part of the zodiac.

Of course the recurrence of similar Mars-phenomena after fifteen years is only approximate, as seven average synodic Mars periods are 19 days short of fifteen years (for more exact figures, see Table 4.7). The Mars loops of successive fifteen-year periods therefore do not fall in exactly the same segment of the zodiac, as can be seen by comparing the loops for 1954 and 1969 at the bottom of Figure 125. In fact, when they are repeated, the loops generally appear in the open spaces left during the first period. Only after 79 years have elapsed with 37 synodic revolutions is a fairly good balance achieved. The Mars phenomena then recur $3\frac{1}{2}$ days later in each case; the loops, too, assume a very similar, though never quite identical, position in the zodiac.

Mars' visibility at different times of year

As the conjunctions and oppositions of Mars shift every two years, its conditions of visibility also change through the seasons. The optimum visibility occurs, on the average, two months later every two years, and so passes through all the seasons over a fifteen-year Mars-period.

Thus the conditions of visibility are considerably different in different years. If the oppositions occur in winter, as for instance in 1990 and 1993 (compare Mars phenomena 1981–2010, Table 6.6), the corresponding Mars-loops will lie in the higher constellations of the zodiac. In the winter months of these years Mars makes a high arc over the southern horizon, and can be seen especially well. It can then be observed much longer than at the time of those oppositions which occur in summer, for instance 1986, 2001 and 2003, when it describes its loops in the lower constellations.

The conditions of visibility thus do not depend on the synodic period alone, but also on the planet's course round the zodiac in just under two years (the sidereal revolution). Over the course of this period the height of culmination rises and sinks between its upper and lower boundaries. This expansion and contraction, however, is interrupted during the retrograde motion, as we have seen for Jupiter and Saturn, by bands in which the arcs are three times as concentrated as elsewhere. As, however, the red planet's synodic period lasts two years and two months, the concentrated bands are not formed in direct succession: more than a whole revolution through the zodiac is completed between them. This means that, beginning at one concentration, the planet must first pass continuously through all the possible heights of culmination, and then complete another two months of regular expansion or contraction before it will again cover a band of the sky three times (Figure 127).

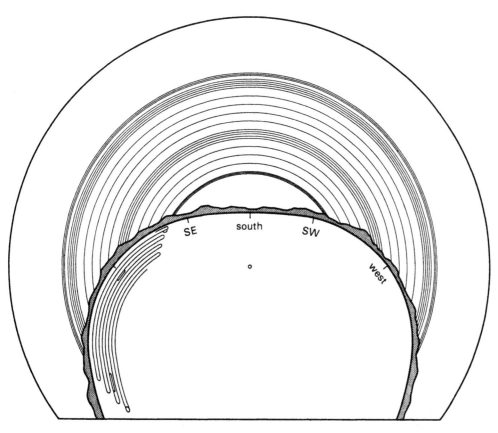

Figure 127. The rhythmic expansion and contraction of the diurnal arcs of Mars during a period of about four years.

Mars' distance from the Earth

The conditions of visibility for Mars, resulting from its higher or lower position in the zodiac, are strongly modified by its changing distance from the Earth. Mars' alternation between perigee and apogee is illustrated in Figure 126 for a fifteen-year period. The overall picture of the planet's movement over such a period shows a strong eccentricity with respect to the Earth, depicted at the centre of the drawing. Mars approaches the Earth in quite a different way in two consecutive loops of opposition. The oppositions which are nearest to the Earth occur at the end of August, when the Sun is in the Lion, and Mars opposite in the Water-Bearer. The most favourable opposition occurs when Mars reaches the ecliptical longitude of 336°. These oppositions are naturally repeated only once in fifteen years. There is an opposition near the Earth on September 24, 1988.

The variations in Mars' distance from the Earth at opposition are considerable.

175

At an extremely close opposition Mars is nearly twice as close to the Earth as at a distant one. The figures are 0.37 and 0.68 astronomical units, respectively. This is dramatically expressed in the planet's luminosity. At a close opposition in the Water-Bearer Mars can have a magnitude of up to −2.8, even outshining Jupiter whose maximum brightness is mag −2.4. At a distant opposition in the Lion, Mars only has a magnitude of −1.1 being outshone by Sirius (at mag −1.6) and all the other planets but Saturn. In both cases, however, Mars is considerably closer to the Earth than the Sun. At conjunction Mars stands behind the Sun, as can be seen in Figure 124, where, for each position of Mars (1', 2' and 3') the corresponding position of the Sun is shown (1, 2 and 3). At its greatest distance from the Earth, Mars is 2.68 AU away, about seven times as far from the Earth as at its closest perigee. Mars thereby displays greater variations in its distance from the Earth than any other planet.

Seen through a telescope, Mars appears as a reddish disc whose diameter varies in size from 3".5 to 26", according to its distance from the Earth.

Seen from the Earth, Mars is fully illuminated by the Sun directly before and after conjunction, and at opposition. Around the time of the quadratures, however, Mars' disc is not fully illuminated, so that it appears gibbous, showing a slight change of phase.

24. Rings and moons of Saturn, Jupiter and Mars

Saturn's rings

Seen through the telescope, Saturn shows a pale yellow-white planetary disc, surrounded by a freely-suspended ring. This gives it its unique character among the planets. It was first observed by Galileo Galilei in 1610, although its ring-character was first recognized by Christian Huygens nearly half a century later.

The apparent diameter of the planet's disc is subject to slight variations. This, together with other astronomical observations (for instance parallax) point to the planet's changing distance from the Earth. Its mean distance is 9.5 astronomical units.

Saturn's disc is not perfectly round, but flattened at the poles. Its surface displays weak darker bands; contoured spots are rare. From the observation of the movement of such spots, however, the planet has been found to rotate with a period of 10¼ hours at the equator. For higher northern and southern latitudes somewhat longer values were obtained. The visible body of Saturn is therefore not a rigid globe. The equator of Saturn is inclined by 28° to the ecliptic, and by 27° to Saturn's orbital plane.

Saturn's ring has three major zones, the centre one being by far the brightest.

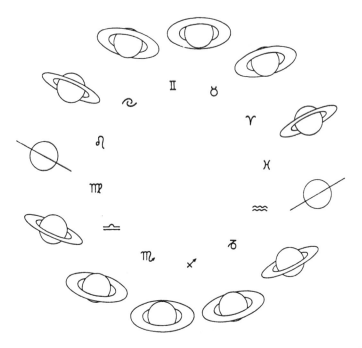

Figure 128. The changing orientation of Saturn's ring to the Earth during a sidereal revolution.

Between the central and outer zone there is a dark line called *Cassini's division*. The division between the central and inner zone is not so pronounced. The transparent innermost ring, called the *crepe ring*, is usually only visible when seen projected against Saturn's disc, where it appears darker than the surrounding area. The recent photographs of Voyager 2 have shown the rings to break up into thousands of ringlets.

The ring rotates in the equatorial plane around the planet. It does not, however, rotate as a rigid unit,* the speed of its motion diminishing with increasing distance from the planet's centre. A point on the innermost ring circles Saturn in only about 5 hours; a point on the outermost edge takes about 14 hours. The inner portions of the ring therefore race ahead of the planet in its 10-hour rotation, while the slower outer portions continually fall behind.

The ring of Saturn appears in different aspects in consecutive years. These positions change periodically in a rhythm of 29½ years, corresponding to Saturn's sidereal revolution (Figure 128).

As the ring-plane is inclined at an angle of 28° to the ecliptic, the upper and

---

*The determination of these movements and speeds was carried out by means of spectral analysis with the help of the Doppler effect.

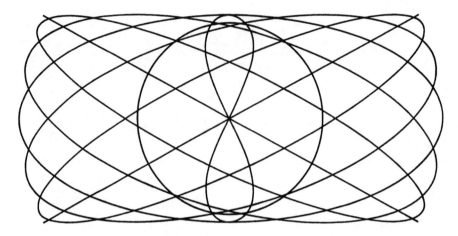

Figure 129. The overall movement of the ring over a thirty-year period.

lower sides alternately become visible. At the transitions the ring-plane coincides with the direction of vision, the edge appearing as a line. Saturn passes through these positions in the constellations of the Lion (169° λ) and the Water-Bearer (349° λ). Similarly, the strongest ring-openings take place in the Scorpion (259° λ) and in the Bull (79° λ). At these times Saturn's disc appears as though embedded in the ring, and completely surrounded by it. Figure 128 illustrates the changing ring-positions in twelve stages. Because of Saturn's varying speed in its passage through the zodiac, the northern side of the ring is visible for a significantly longer time (16 years) than the southern side (13½ years).

The changing position of the ring causes a considerable variation in Saturn's total luminosity, which is even noticeable for naked eye observation. During its 29½ year revolution, therefore, the planet undergoes a double increase and decrease in its brilliance. Its maximum brightness naturally coincides with the time of widest ring-opening; the minimum takes place when the ring is invisible. The difference is equivalent to a magnitude of about 1.0, meaning that Saturn is nearly twice as brilliant during the greatest ring opening than during the smallest opening. Table 6.8 of the Appendix shows the years of extreme ring orientation.

The transition through ring-invisibility can take place under different conditions, depending on Saturn's position within its synodic period. If Saturn is in opposition to the Sun during the transition, so that Saturn, Earth and the Sun are in a line, only the ring's edge is illuminated, causing it to remain practically invisible. The perspective is similar at conjunction, although the transition is then altogether invisible. If Saturn is in quadrature during the transition, one side of the ring is illumined, while the other is visible before or after the transition as a dark band across the planet's disc. Naturally, all intermediate stages between these two extreme conditions also take place.

178

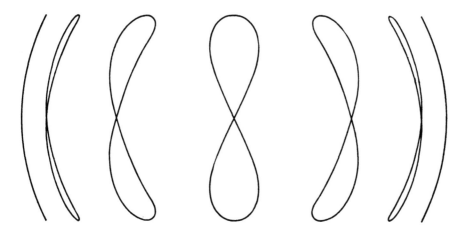

Figure 130. Lemniscatory movement of individual points on the ring in thirty years.

Twice every sidereal revolution one or three successive transitions through ring-invisibility take place. 1979 has a single transition, whereas a triple transition takes place in 1995. In 1936/37 a rare situation occurred when two of the three transitions coincided as the ring first became invisible.

A survey of the movements of Saturn's rings with respect to the Earth is given in Figure 129 for a sidereal revolution. The changing position of the rings along with the varying breadth of the opening and the degree of tilt is shown in pictorial form.

The overall movement of the ring — disregarding the additional slight variations from year to year — shows the same general character of a 'scissoring' up and down movement, which we have already seen for various related phenomena, such as the annual displacement of the Sun's equator relative to the Earth.

If we fix our attention on the mid-point of the edge, we find that it describes a spherical lemniscate in its rising and sinking movement over the thirty-year period (Figure 129). Points on the ring edge which are separated by 30° and 60° from the mid-point describe the lemniscates shown in Figure 130.

Saturn's moons
Apart from its ring Saturn is circled by nine named moons, but the Voyagers discovered several too small to be seen from Earth, bringing the total to over 20. The four nearest named moons to Saturn move around the central body in extremely short periods: 18 hours, $22\frac{1}{2}$ hours, 1.4 days and 1.9 days. The more distant satellites have periods of revolution of several days and weeks; the outermost, Phoebe, requires $550\frac{1}{2}$ days, and shows yet another abnormal trait: it circles the planet in reversed direction, that is retrograde, and thereby falls out of the uniform system of the remaining moons.

Jupiter's moons

Jupiter seen through the telescope appears as a bright yellowish-white planetary disc. Its apparent diameter is also subject to slight variations (30″ to 50″). From these and from angular measurements a mean distance of 5.2 AU has been calculated.

Obliquely across the planet a row of darker bands of unequal breadth can be seen. These are parallel to Jupiter's own rotation, which can be observed from the surface details, and in particular, from the so-called Great Red Spot. It is found to be slightly less than 10 hours ($9^h 50^m$), but decreases slightly ($9^h 55^m$) towards the poles, which show a fairly strong flattening. The planet's axis of rotation is nearly vertical to its orbital plane. Almost exactly in the plane of Jupiter's equator (which nearly coincides with the ecliptic) circle four large moons, known as the Galilean satellites. They were observed for the first time in the years 1609/10 by Galileo Galilei in Florence and by Simon Marius (Mayr), an earlier pupil of Tycho Brahe, in Prague, who were the first to apply the telescope to astronomical observation. These four bright moons circle their planet at distances of 3 to 13 Jupiter-diameters, and with periods of 1.8, 3.6, 7.2 and 16.8 days. At the end of the last and in the course of the present century a number of further Jupiter moons were discovered. These satellites (at least sixteen) are all very small and dim. The eight outermost moons have strongly eccentric orbits as well as being strongly tilted to Jupiter's equator. Like Saturn's outermost moon, four of Jupiter's (XII, XI, VIII and IX) are retrograde.

The Great Red Spot

Jupiter's Great Red Spot, although not quite so prominent telescopically as Saturn's ring, is an equally unique and interesting feature. It appears as a broad oval area in the planet's southern hemisphere, whose length is equivalent to about a quarter of the planet's diameter. Its colour, normally pale pink or mauve, as well as its brightness, vary at irregular intervals. In 1713 it faded altogether, and was not observed for about 120 years. In the period from 1878 to 1882 it attained such an intense reddish colouration that it could be distinguished with a pair of binoculars. This intensity was reached again only in 1936. This seems the more remarkable when we consider that the spot had been known for over two hundred years since its discovery by Cassini (1665) before its red colour was first observed by Lawrence Parsons, Earl of Rosse, in 1872 through his 72-inch reflector. Such periods of unusual brightness are normally followed or preceded by a great weakening or virtual disappearance, as was the case in the years following 1882, and again in 1926 and 1936.

Mars' satellites

Mars appears through the telescope as a reddish disc. The considerable variations of its diameter, its distance from the Earth, and its slight change of phase have already been discussed in Chapter 23. The slight flattening of Mars at the poles is a little less than that of the Earth.

The surface-features of Mars are well known through photographs transmitted by space-probes. These reveal numerous craters and mountain-ranges not unlike those of the Moon, as well as complex surface-patterns which can only be understood as the effect of water-erosion. These details cannot be satisfactorily resolved in the telescope, where they appear as dark bands and points separating brighter regions. Among the variable spots two bright polar areas are especially notable. Their girth increases and decreases, according to whether 'summer' or 'winter' prevails on the corresponding Mars-hemisphere.

From these surface features Mars' rotation about its own axis has been determined exactly at 24^h 37^m. The inclination of the planet's equator to its orbit is 24°, similar to that of the Earth. The seasons, however, are not, as we would expect, of roughly equal length. The orbit of Mars around the Sun is so eccentric that the 'summer' for its northern hemisphere lasts 371 Martian days, whereas the 'winter' lasts only 298 days.

Strong telescopic magnification under good observational conditions reveals two small Mars-satellites, the moons Phobos and Deimos (Fear and Fright). These rather discomforting names were borrowed from the *Iliad* (Book XV) where Mars (Ares) is described preparing to descend to the battle-field to avenge the death of one of his sons:

And he ordered Phobos and Deimos to harness his horses,
While he himself donned his sparkling armour.

The moons were discovered in 1877 during a favourable Mars-opposition. They circle the planet extremely closely and at very high speeds. Phobos is only 1.4 Mars-diameters from the centre of the planet, and completes a revolution in a mere 7^h 38^m, that is, about three times faster than Mars rotates around its own axis. The second moon, Deimos, circles the planet once in about 30 hours at a distance of 3.5 Mars-diameters.

25. The loop-formation of Saturn, Jupiter and Mars

The formation of loops is a feature which strongly distinguishes the planetary motions from those of Sun and Moon.

A few introductory remarks may be allowed before we discuss these movements in detail. Ever since the old star-wisdom of the Greek mysteries evolved to a science, the most recurring and challenging problem was the geometrical representation and interpretation of the planets, in their enigmatic and tortuous wanderings. The planetary loops served as stimulus and inducement for the astronomical labours to which we owe such important ancient theories as those of Eudoxus and Ptolemy. At the dawn of the modern age Copernicus and Kepler offered new solutions to these ancient riddles.

Even among astronomers there are few who are fully conversant with the loops of the different planets in their specific details and in their rich differentiation. The reason for this lies in the fact that a closer study of these phenomena seems superfluous, since the Copernican heliocentric theory has demonstrated the illusory character of the loops. This theory showed that the latter do not belong to the 'real' planetary movements, as was naïvely assumed in the Middle Ages. The planetary loops were shown to be projections of the planets' movements, resulting from the simultaneous but unequal motion of the Earth and the planet along nearly circular (elliptical) orbits around the Sun.

In the heliocentric system the apparent planetary movements are written off as relative movements, depriving them of much of the significance ascribed to them in ancient times. If, however, we widen the scope of the question, their importance can be rediscovered on a new and higher level. For a *relation* must lie at the basis of any movement; and this is to be sought in the fact that the motion of the Earth and of a particular planet stand spatially and temporally in a certain measurable relationship.

While the loop-orbits do not in any naïve sense show the absolute movement of the planet as such, they give a picture of the specific relative movement of the Earth and the planet. In the nature of the relationship of two things to each other something is always reflected of their *behaviour*. This can be experienced at first hand in the study of the loop-movements. Something of the 'behaviour' of each planet towards the Earth is expressed in its own unique way in the manner in which its loops are disposed and distributed in the sky.

Our method of representation, which attempts to proceed in Goethe's sense, may again be briefly summarized. Certain single astronomical phenomena are brought together, and further phenomena resulting from these associations are followed. If our survey of the sequences in space and time is carried far enough in this manner, certain comprehensive *pictures* arise naturally out of the single phenomena, each of which makes its own contribution to their formation. Here we find the possibility of a further step, if these pictures for different planets are compared. Such comparative contemplation can make visible *archetypal phenomena*.* In becoming aware of these, we gain access inwardly to an ordering principle which prevails unseen in the outward course of events from which we began our investigations.

*Goethe's *Urphänomene*

182

25. THE LOOP-FORMATION OF SATURN, JUPITER AND MARS

The nodes

The individual loops of a single planet vary their forms, just as do the loops of the different planets among one another. It will be our concern to find the principles which govern them.

Exact observation shows that Saturn, Jupiter and Mars stand north of the ecliptic during one half of their revolution through the zodiac, and south of the ecliptic during the other half. At two opposite points in the zodiac their orbits intersect the ecliptic; these points are called, as for Moon, Venus and Mercury, the northern (Ω) and southern ($\mho$) nodes. For Saturn and Jupiter they are located in the constellations of the Twins and the Archer, although for Saturn they lie rather more in the direction of the Crab and the Goat. (Saturn Ω 113°, $\mho$ 293° λ; Jupiter Ω 100°, $\mho$ 280° λ). Mars' ascending node lies at the end of the Ram (49°), the descending node in the Scales (229°). Like the lunar nodes, the planetary nodes are not fixed. However, their displacement occurs extremely slowly, and can be disregarded here.

The inclination of Saturn's orbit to the ecliptic is roughly twice that of Jupiter's. Saturn's greatest northern and southern divergence from the ecliptic is 2°5; Jupiter's is about 1°3. Mars lies between them with 1°85. These maximum latitudes are always reached halfway between the two nodes: that is for Saturn in the Virgin and the Fishes (203° and 23° λ), for Jupiter in the same constellations, but at 190° and 10° λ, and for Mars at the beginning of the Lion and in the Goat (139° and 319° λ).

The shape of the loops

The loops which are described at the nodes of the planetary orbits are open, S or Z-shaped loops, with the slanting retrograde portion always crossing the ecliptic. At intervals of 90° from the nodes the loops are closed and symmetrical. In the intermediate positions transitional forms are described.

It will be seen that the same principle underlies the loops of the outer planets as we have already seen for the inner planets Venus and Mercury. The difference lies in the fact that the heliocentric orbits of the latter two planets can be imagined situated before us, at any given moment, over a relatively small segment of the sky. Their particular orientation during loop-formation makes the individual loop-forms easy to comprehend. These orientations, however, are in turn dependent on the position of the Sun with respect to the planet's nodes. In Figure 115 positions c and g occur when the Sun is situated at the southern and northern nodes of Mercury, respectively. For the outer planets the loops come about as it were in response to the Sun's movement on the opposite side of the Earth, which we have described as a kind of 'resonance'. In general, we can say that the loops of the inner planets arise directly out of the progression of their heliocentric orbits through the zodiac, while those of the outer planets are a projection of the Sun's movement around the Earth. This fundamental polarity was known to the ancient

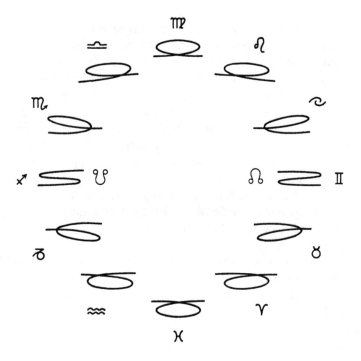

Figure 131. Jupiter's loop metamorphosis in the zodiac (the underlying principle is emphasized).

Greek astronomer, Ptolemy and was applied in his system of epicycles, by means of which he attempted to reduce the loop-motions to a complex of regular circular movements.

Jupiter's loops

Let us now consider the loop-forms in more detail, using Jupiter as an example. Figure 131 shows twelve Jupiter-loops schematically, as they arise in the zodiac during one sidereal revolution. To facilitate the comparison, they have been turned into the horizontal.

The loop in the Virgin is symmetrical and is turned upwards (northwards); the loop in the Fishes is correspondingly turned downwards (southwards). At the nodes in the Archer and the Twins the S and Z-forms are described, respectively. Two transitional forms lie between each pair of these extremes. In its progression from the Virgin to the Scales, the northward-directed loop loses its symmetry; in passing from the Scorpion to the Archer it opens to the S-form. It then closes again on the other side, and aspires out of its asymmetrical form to the southward-directed symmetry in the Fishes. The point of intersection is now displaced to the right, until the loop opens once more in the Twins. It closes, finally, towards the

184

north, and reaches the symmetrical form of its starting-point in the Virgin. This transformation can be regarded as a kind of cosmic archetype of metamorphosis.

Moreover, an axial symmetry will be found between the right and left-hand side of Figure 131, and another symmetry between the upper and lower halves. A closer study will reveal still further symmetries. In the transformations, therefore, the most manifold mutual relationships are contained, whereby one form counterbalances the other.

As a loop is displaced over longer intervals of time, all the others must follow it. But a new 'inner equilibrium' of the individual components is attained. Alongside the archetype of metamorphosis, arises the picture of a kind of great loop-organism, whose individual members are mutually and harmoniously accommodated to the whole.

Figure 119 illustrates this whole transformation in its proper orientation and scale within the zodiac for one sidereal revolution (1959–1970).

Saturn's loops

For Saturn this organism contains twenty-eight loops, whereby the progressions from one loop-form to the next are smaller. The fundamental principle, however, is the same (compare Figure 122).

Mars' loops

The loop-organism of Mars, in its first stage, encompasses seven to eight loops in fifteen years. Figure 132 shows the form-sequence in the circle of loops, with the appropriate tilt to the ecliptic. In the Crab and the Lion, Mars' loop attains its greatest northern latitude above the ecliptic; in the Goat and the Water-Bearer it is furthest south of the ecliptic. At the nodes the open S or Z-forms are described,

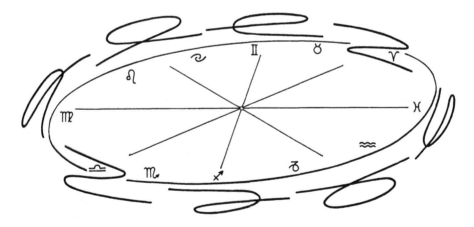

Figure 132. The forms of the loops of Mars (schematically).

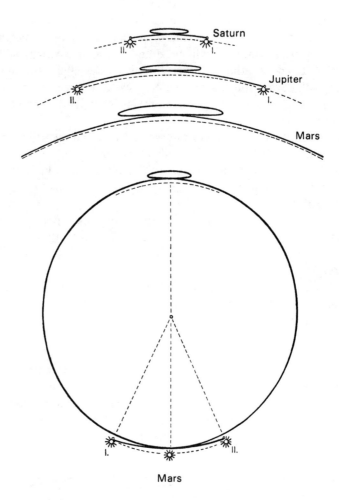

Figure 133. A comparison of the loop-orbits of the outer planets during one synodic period.

but in this case only one transitional form is present. On the whole, we can recognize the same polarities and symmetries as for Jupiter.

In the following fifteen-year period the loops have shifted roughly to the open spaces between the forms in Figure 132. All the loop-forms thereby pass through intermediate stages to those illustrated. The mutual relationships, however, are retained. Only after a long series of such transformations, is a situation finally attained which is very similar to the starting-point.

Each year Saturn and Jupiter display a new form in the sequence of their loop-organism; Mars, on the other hand, describes a loop only every second year, also as the next in the series. But in the meantime it wanders through the whole of the zodiac, as we have already seen.

186

25. THE LOOP-FORMATION OF SATURN, JUPITER AND MARS

Loop size

Let us next consider the size of the loop. The length of the retrograde motion increases from Saturn (6.5) over Jupiter (10°) to Mars (average 15°); its duration, however, decreases (Saturn averages 140 days, Jupiter 120^d, Mars 74^d). Also in relation to the length of the whole synodic period, the loops diminish in importance from Saturn to Mars.

The upper part of Figure 133 shows the relative sizes of the loops of Saturn, Jupiter and Mars, giving the direct movement and the average loop-size. The return of a conjunction with the Sun (synodic period) is indicated by points I and II. For Saturn and Jupiter this requires slightly more than a year. Mars' synodic period, which lasts over two years, is represented in the lower part of the figure.

Further characteristic differences among the three planets can be found by comparing the ratio of direct to retrograde motion; the average values are illustrated schematically in Figure 134. Saturn moves a good 18° direct and 7° retrograde. This gives a ratio of about 3:1, that is, between the loops lies a portion of direct orbit which is roughly as long as a loop. Or, to put it differently, Saturn advances three steps, and then goes back one step, describing the retrograde part of a loop.

For Jupiter the mean direct motion is 43° and the retrograde motion on the average 10°; it is roughly the ratio of 4:1. The direct motion between the ends of two loops is twice as long as a loop. Accordingly, we can say: Jupiter takes 4 steps forwards, 1 back, and so on.

The corresponding values for Mars are subject to great variations. The average values can be taken as 410° for the direct motion and 15° for the retrograde motion. This gives a relationship of about 27:1.

Jupiter describes a succession of medium-sized loops in harmonious proportion. Saturn's small, narrow loops are arranged in closer sequence. For Mars, by contrast, the larger, broadly swung loops seem almost incidental when compared to its two-year revolution through the zodiac.

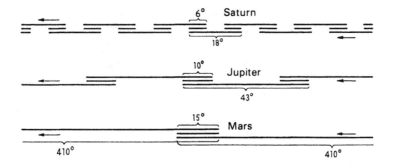

Figure 134. The relationship of direct to retrograde motion for Saturn, Jupiter and Mars.

The expansion and contraction of the diurnal arcs of the planets and their synodic periods are intimately connected with the phenomena we have just described. These themes have been treated separately for each planet, and may now briefly be recalled in passing. *Jupiter's* expansion from the smallest to the largest arc is articulated into five or six stages of contraction. Correspondingly, five or six expanding stages interrupt the planet's contracting course from the Twins to the Archer. *Saturn* crowds together more than double the number of stages; there are fourteen or fifteen in the expanding phase, and the same number in the contraction. *Mars*, on the other hand, covers its whole celestial field more than once between two contractions. Here, too, we see the fundamentally different character of Mars from the related movements of Jupiter and Saturn.

The rhythmic subdivision of the three planets' celestial fields is an expression of the characteristic relationship of their revolution around the zodiac to their synodic period. *Saturn* takes 29.5 years to wander around the zodiac, but has the relatively short synodic period of 378 days. *Mars'* sidereal revolution is about 3 months shorter than its synodic of 780 days. *Jupiter*, which requires roughly 12 years to circuit the zodiac, and 399 days to complete a synodic period, stands between the two in this respect, although very much closer to Saturn.

Colour and brightness

Let us finally look at the variation in brightness of Saturn, Jupiter and Mars. The fixed stars, with a few exceptions, always retain the same brightness and class of magnitude. They radiate their own light. The planets, however, reflect the sunlight, although somewhat modified in each case. This comes to expression in the various shades of colour: Saturn's pale yellow-white, Jupiter's yellow, Mars' reddish-orange. These various tints show us that the planets do not pass on the sunlight exactly as they receive it. While the colours are qualitative properties of the planets, we can also use a quantitative indicator, called the *albedo*. The albedo expresses the ratio of the light which a planet reflects in all directions to the light received. The albedi of Jupiter and Saturn are found to be 0.70 and 0.75, respectively, meaning that they reflect 70% and 75% of the sunlight. Mars, on the other hand, retains all but 16%, that is, it has an albedo of 0.16. The Moon has a still lower albedo of 0.07, making it one of the least reflective bodies in the solar system.

The apparent brightness of the three outer planets depends on the one hand on their angle from the Sun, and on the other hand, on their proximity to the Earth. (The relatively slight fluctuations which result from the planets' changing distance from the Sun need not be treated here). If the Sun's glare could effectively be eliminated during the time of a conjunction with one of these planets (which is practically the case during a total eclipse), the planet would be visible in the Sun's neighbourhood. Because it is behind the Sun it does not undergo an interruption of its brightness as does the Moon at its new phase.

Figure 135 shows the variations in brightness for Saturn, Jupiter and Mars over

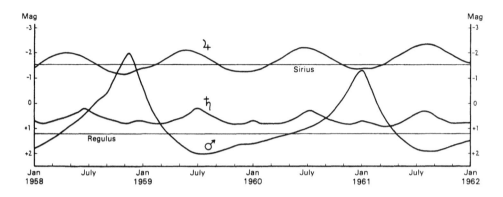

Figure 135. Average variations in brightness for Saturn, Jupiter and Mars. For comparison, the magnitude of Regulus and Sirius is shown.

a four-year period. (By way of comparison, the brightness of Regulus in the Lion and of Sirius are given.) The peaks of these light-curves are attained at opposition to the Sun, and succeed each other at intervals of one synodic period.

Saturn's brightness varies by an average of about 0.5 classes of magnitude during its revolution. Through the additional factor of the changing position of its rings, however, its total range of variation is from mag 0.02 to 1.5. Jupiter's magnitude varies from −2.0 to −2.4 during the opposition; were it visible at conjunction it would have a magnitude of −1.2 to −1.5. The average variation is about 0.8 classes of magnitude. We have already mentioned Mars' huge variation of 4 classes. As with all the Mars phenomena, however, considerable differences are possible here. The greatest brightness which the planet attains is −2.8; it is reached during oppositions which are especially near the Earth. In those oppositions which are farthest from the Earth its magnitude is only −1.0. Mars' ascent to its greatest brightness and its subsequent decline take place very quickly.

On the whole, Saturn shows the most tranquil light-curve, that is, the smallest variations. Jupiter's variations are slightly greater, while those of Mars are most dramatic and extreme.

From whatever point of view we may compare the phenomena of the three outer planets, Mars always occupies an extreme position which is not easy to resolve with the more tempered and harmonious phenomena of Saturn and Jupiter. This may be seen as the geocentric aspect of a problem which was for many years a puzzle to astronomers who contemplated the heliocentric arrangement of the planets. In 1766 Titius, professor in Wittenberg, published a mathematical progression which expressed in good approximation the average distances of the planets from the Sun. Strikingly, one member of the series was found to be missing between the orbits of Mars and Jupiter. This disproportion, to which Kepler had

189

already pointed nearly two centuries earlier, led Bode, the director of the Berlin Observatory, to assign an association of 24 astronomers with the task of seeking the missing planet. Not until the beginning of the nineteenth century, however, was the gap to be filled by the discovery of a number of minor planets, or planetoids. We shall consider these in more detail in Chapter 28.

26. A comparison of the superior and inferior planets

The superior planets Saturn, Jupiter and Mars attain their best visibility when they can be seen throughout the night while in opposition to the Sun; they are markedly 'night-planets' (Figure 136). The inferior planets, Venus and Mercury, on the other hand, are most often above the horizon during the daytime: they are predominantly 'day-planets'. Their visibility is confined alternately to the evening and the morning (Figure 137).

The outer group passes through a time of visibility and a time of invisibility in

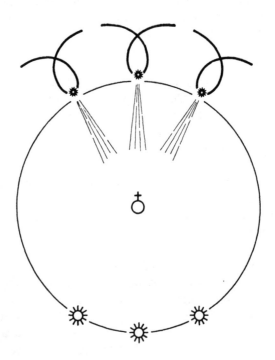

Figure 136. The optimal visibility of Saturn, Jupiter, and Mars at their opposition to the Sun.

190

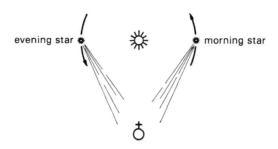

Figure 137. Optimal visibility of Venus and Mercury as evening and morning star.

the course of each synodic period. The synodic periods of the inner group are divided into two periods of visibility and two periods of invisibility, which come to expression in the rhythmic alternation of the evening and morning periods (Figure 138).

Saturn, Jupiter and Mars become visible in the morning sky after their conjunction with the Sun; at opposition they reach their best visibility and maximum brightness, and then sink slowly into the evening twilight, whereupon they come once more into conjunction with the Sun. At opposition they stand in perigee, at conjunction in apogee, but always beyond the Sun. For this reason they are known as the outer, or superior planets.

Venus and Mercury alternate between superior and inferior conjunctions; in between lie the positions of maximum brightness in the evening and morning sky, relatively close to the horizon. They are never visible at midnight, nor do they ever come into opposition to the Sun. At superior conjunction they stand in apogee beyond the Sun; at inferior conjunction in perigee between Sun and Earth. They are therefore known as the inner, or inferior planets.

In Chapters 22 and 23 we spoke of the expansion and contraction of the daily arcs of movement, and of their concentration during loops of Saturn, Jupiter and Mars. Because Venus and Mercury can only be observed in the evening and morning sky, their diurnal arcs do not appear in their totality; only the portions near the horizon can be followed with the unaided eye. Together with the planets' movement through the zodiac, these arcs wander towards the north or the south along the western and eastern horizons. For the superior planets the threefold sweep over a band of the sky takes place during the maximum brightness, at opposition to the Sun; for the inferior planets the concentration is woven about the Sun at a time of invisibility.

In successive years the phenomena of the superior planets are always repeated later (Saturn 13[d], Jupiter 34[d] and Mars 50[d]); those of the inferior planets are repeated earlier (Venus 5 months, Mercury 18[d]).

In their movement through the zodiac the superior planets are invariably

191

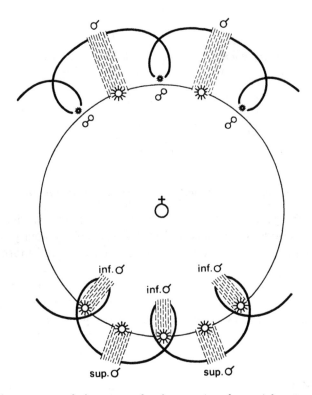

Figure 138. The sequence of phenomena for the superior planets (above) and the inferior planets (below). Where the orbits are interrupted by the rays of the Sun, the planets are invisible.

overtaken by the Sun; while the inferior planets alternately overtake the Sun and are overtaken by it.

A fundamental polarity between the superior and inferior planets is brought to light by a consideration of their loops. For Saturn, Jupiter and Mars the emphasis on the loop-principle in the planets' overall movement is an important characteristic. For Venus and Mercury the loop-formation is practically invisible; in its stead, the radial principle in the planets' movement towards and away from the Earth takes on a special significance (Figure 138). In this respect, Mars assumes a kind of intermediate position. Its radial movement towards and away from the Earth plays a stronger role than for any other planet. Its loops, at the same time, are less frequent than those of any other planet.

The inclination of the orbit to the ecliptic and the directions of the nodes are different for each planet. Owing to the different positions of the nodes, the basically similar cycle of loop-forms is variously orientated with respect to the zodiac. The *S* and *Z*-formed loops of Mercury and Mars, for example, are restricted to the constellations of the Ram and the Scales (Figures 98 and 125); those of

192

Venus fall in the direction Bull-Scorpion (Figure 83), and those of Jupiter and Saturn in the direction Twins-Archer (Figure 119 and 122).

Another particularly characteristic difference between the two groups of planets lies in the changing phases of the one and the moons of the other. Venus and Mercury are, as we have seen, to a certain extent 'lunar' in character. In this respect they form a single group together with the Earth's Moon. All three luminaries are able to pass directly in front of the Sun and thereby to eclipse a greater or lesser part of its light. Mars in quadrature to the Sun is slightly gibbous, showing a small change of phase; Saturn and Jupiter exhibit no phase whatever. The three superior planets, however, are encircled by a multiplicity of moons; Saturn has in addition a ring. From the Earth they can never be seen in front of the Sun, but their satellites can pass across the planetary discs.

In conclusion, let us look briefly at the magnitude graphs (Figures 116 and 135). The variations in brightness over the course of a synodic period increase considerably from Saturn over Jupiter to Mars. Venus attains, on one hand, the greatest luminosity of all the planets; on the other, its light is extinguished almost totally when it stands before the Sun. The curves of Venus and Mercury are interrupted at each inferior conjunction. Exceptions to this are possible, and even characteristic, for Venus. The Moon assumes a natural place within this group; each year it alternates twelve times between maximum brightness and darkness (Full and New Moon). Among the inferior planets Venus inclines in character towards the superior ones; the contrary is the case for Mars.

The magnitude graphs fluctuate increasingly from Saturn to the Moon; the element of darkness enters more and more incisively. From the inmost to the outermost planets we find an increasingly harmonious and steadfast quality in the light phenomena, which emerges with growing sovereignty.

All these manifold fluctuations of the planets' luminosity take place between the tranquil radiance of the fixed stars and the dark, night-enshrouded Earth.

27. Uranus, Neptune and Pluto

Since the eighteenth century, telescopic observation has disclosed the existence of three further planets not immediately visible to the unaided eye: Uranus, discovered in 1781; Neptune in 1846; and Pluto in 1930. Uranus, whose brightness increases to about magnitude 5.7, stands just on the threshold of visibility. Neptune and Pluto, however, with maximum magnitudes of 8 and 14, respectively, can only be observed with a telescope.

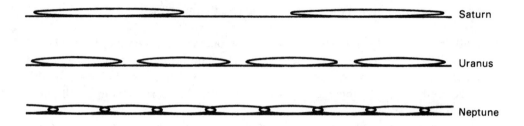

Figure 139. Comparison of the loops of Saturn, Uranus and Neptune.

The movements of these three outermost planets are roughly analagous to those of Jupiter and Saturn. Their synodic periods, however, and accordingly their whole cycle of visibility and invisibility, are only very slightly longer than a year. Uranus requires 1 year $4\frac{1}{2}$ days; Neptune, 1 year $2\frac{1}{4}$ days; and Pluto, 1 year $1\frac{1}{2}$ days. During each synodic period, the planets' normal movement through the zodiac is interrupted by a period of retrograde motion, which results in the formation of a loop. In conformity with the slower movement of these planets, the loops are very small. Those of Uranus measure about 4° and are a little more than half as long as Saturn's loops. Neptune and Pluto describe loops of almost 3° and 1°, respectively. Moreover, a comparison of the individual movements shows that the loops are traced closer and closer to each other as we pass from the more inward to the more outward planets. Those of Neptune and Pluto even overlap to some extent; this leads to the interesting consequence that both planets pass through certain small portions of the zodiac five times in the course of two years (Figure 139).

Uranus, Neptune and Pluto all have long sidereal revolutions, roughly proportional to their increasing distances from the Sun. Uranus requires 84 years to pass through the whole zodiac; Neptune, 165; and Pluto, 248.

In consequence of their long periods of revolution, the three planets advance only a few degrees each year. Uranus progresses $4°3$, Neptune $2°2$, and Pluto $1°5$ annually. Uranus requires seven (7.00) years to transverse a twelfth of the zodiac, or 30°; Neptune takes fourteen years (13.73) and Pluto twenty-one years (20.64). From these proportions it can be seen that the sidereal revolutions of the three planets stand almost exactly in a relationship of 1:2:3.

At present (1986) Uranus is in the constellation of the Scorpion, moving towards the Archer, Neptune in the Archer, and Pluto in the Scales.

The orbits of Uranus and Neptune are inclined to the ecliptic by only $\frac{3}{4}$° and $1\frac{3}{4}$°, respectively. The nodes of Uranus lie in the Bull (Ω) and the Scorpion ($\mho$), those of Neptune in the Crab (Ω) and the Goat ($\mho$), and those of Pluto in the Twins (Ω) and the Archer ($\mho$). Unlike those of Uranus and Neptune, Pluto's orbit crosses the ecliptic at the unusually steep inclination of 17°. In this respect

Pluto assumes an exceptional position among all the planets, with which only Mercury, with its orbital inclination of 7°, invites comparison. In its most extreme declinations in the Virgin and the Fishes, Pluto moves along the outermost edge of the band of zodiacal constellations.

Satellites

Viewed through the telescope, the small planetary disc of Uranus appears in bluish or greenish tinges. Neptune's pale light has bluish nuances; its outline is so vague that it is difficult to distinguish from a fixed star. Pluto appears as a yellowish point of light.

Five moons have been discovered around Uranus. They move in almost circular orbits, and all in the plane of the planet's equator. The moons require only a few days for their revolutions (1.4, 2.5, 4.1, 8.7 and 13.5 days). Recent observations show Uranus to have a number of narrow rings.

Uranus itself rotates around its own axis in only $10\frac{3}{4}$ hours. The most unique feature of Uranus' motions is that its direction of rotation is nearly perpendicular to its orbit around the Sun. Its equatorial plane, in which the moons encircle the planet, is inclined at 82° to its orbit. Its direction of rotation is also opposite to that of the other planets. For this reason astronomical tables give the supplementary angle of inclination: 98°.

When Uranus stands in the constellations Scorpion (1902 and 1986) and Bull (1946) our glance falls nearly perpendicularly onto the orbit of the moons and the one pole of the planet. When it stands in the Water-Bearer (1924 and 2008) and the Lion (1966) we see the orbital plane on edge.

Uranus exhibits in its sidereal revolution an unusually harmonious relationship to the Sun. Its orbit deviates less strongly from the ecliptic than that of any other planet. With respect to its more individual motions, namely the rotation around its own axis and the revolution of its moons, it appears to be almost completely emancipated from the prevailing patterns of the solar system.

Neptune is known to possess two moons. Triton, which circles the planet in slightly less than 6 days, has an orbital inclination of 160° (or 20° retrograde) to Neptune's equator, so that its retrograde motion is even more striking than that of Jupiter's moons. The second satellite, Nereid, is very much smaller than Triton, and possesses a large and strongly eccentric orbit. Nereid requires nearly a year to complete a single revolution around the planet. Its orbital inclination is slightly greater than that of Triton, though the direction of its motion is forward. Neptune's own equator is inclined at almost 29° to its orbit. The rotation is also forward, and takes place in about 18 hours.

Pluto appears so small even in the largest telescope, that no distinct surface features can be recognized. Certain periodic variations in brightness, which are presumably due to a different reflectivity (albedo) on different portions of the planet's surface indicate a rotation period as 6.4 days. In 1978 astronomers disco-

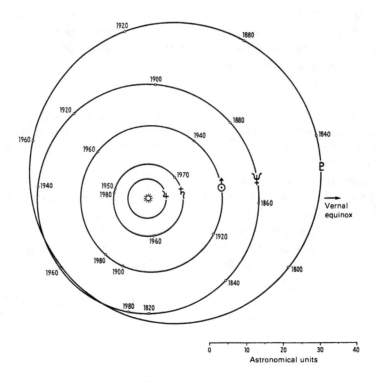

Figure 140. The heliocentric orbits of Jupiter, Saturn, Uranus, Neptune and Pluto. The arrow points to the vernal equinox.

vered that Pluto has a moon, named Charon, in size about 40% of Pluto's diameter. It orbits Pluto in the same period as the planet's orbital rotation.

Pluto's distance from the Earth and the Sun is subject to unusually large variations. These relationships can most easily be grasped by comparing the calculated orbits of the outer planets around the Sun, as they are illustrated in Figure 140. Those of Uranus and Neptune form almost concentric circles, with the Sun near the centre; the path of Pluto is a strongly eccentric ellipse. In the years of its perihelion (which, owing to its great distance, are the same as its year of perigee) Pluto is even nearer the Sun than Neptune. Pluto stands in perihelion in 1989 in the constellation of the Scales (224°). Its aphelion (in the Ram, 44°) was last attained in the year 1865. In consequence of the strong inclination of Pluto's orbit, the planet does not directly cross Neptune's orbit as it approaches aphelion, but passes considerably above (north of) the latter.

The calculation of the relative distances of these 'outsiders' of the solar system leads to the following results: the mean distance of Uranus from the Sun is 19.2 astronomical units; that of Neptune is 30.1 AU; and of Pluto, 39.4 AU (on account of its eccentric orbit it varies between 29.6 AU at perihelion, and 49.3 AU at

196

aphelion. The three planets thus succeed one another at intervals roughly equal to the distance between Saturn and the Sun (9.5 AU). Uranus therefore stands at double, Neptune at triple, and Pluto, at four times the distance of Saturn to the Sun. (The planets' mean distances from the Earth are, of course, identical to their mean distances from the Sun.)

Discovery of the outer planets
Uranus was first observed by chance by the astronomer Sir William Herschel, as the planet came into the field of vision of his reflector-telescope on March 13, 1781. Herschel noticed a small disc in the constellation of the Twins, whose diameter increased with increasing magnification. He was soon able to distinguish its movement. It was Laplace, however, who was first able to demonstrate that this was a newly-discovered planet with a calculable orbit around the Sun.

The discovery of Neptune took place of September 23/24, 1846 by J. G. Galle in Berlin, on the basis of indications of the French astronomer Urbain Leverrier for the position of a further, long-suspected planet beyond Uranus.

For several decades it had been known that the observed positions of Uranus deviated from those expected on the basis of calculations. The deviations became stronger in the period between 1830 and 1846, and the assumption became inescapable that they stood in connection with the disturbing influence of an unknown planet beyond the orbit of Uranus. In 1845, almost simultaneously, and completely independently, Leverrier in France and J. C. Adams in England calculated the supposed position of the new planet within an accuracy of one degree.

Galle's rapid discovery of the faint planet on the same evening as the arrival of Leverrier's letter indicating its calculated position, was only possible with the help of a new and very exact map of the same portion of the sky, which had just been published. Near the predicted position, between the constellations of the Goat and the Water-Bearer, Galle noticed an object of the eighth magnitude which did not appear on the map. On the following days the star in fact showed a very slight displacement against the fixed-star background, confirming its planetary character.

The planet Pluto was discovered at the end of January, 1930 at Lowell Observatory in Flagstaff (Arizona). It appeared as an extremely faint point of light of fifteenth magnitude in the constellation of the Twins. On the basis of his calculations of disturbances in the orbit of Neptune, Lowell had, as early as 1905, expressed the possibility that there might be up to three planets beyond Neptune. He had calculated their orbits, but all attempts to confirm their existence through observation remained fruitless for many years. The discovery in 1930 took place in fact not far (5°) from the position which Lowell had calculated.

Uranus, Neptune and Pluto form a group of their own insofar as they remain invisible to the unaided eye. Their kinship can, however, also be sought in certain deviations from the order of the rest of the solar system: in the inclinations in part of their own axes of rotation, in part of the orbital planes of their moons, and in the resulting retrograde motions. (The latter is also true of certain moons of Jupiter and Saturn; here, however, it is strongly outweighed by the 'normal' rotation of the majority of the satellites of these two planets).

In comparing the phenomena of the three outermost planets we find a number of striking similarities between Uranus and Neptune, while Pluto stands as a kind of outsider. It appears significant that the irregularities in the motions of Uranus and Neptune are not manifested directly in the planets' sidereal revolutions through the zodiac. On the contrary, both planets display unusually regular motions with respect to the Sun. Their orbital inclinations of 0° 46' and 1°46', respectively, are among the smallest deviations from the ecliptic of all the planets in the solar system. This results in a corresponding flattening of the planets' loops, particularly in the case of Uranus. The metamorphosis which we observed above all in the loops of Mars, Venus and Mercury, almost disappears. The impression arises of 'frozen' clockwork movements lacking the vitality of the closer planets. This impression is reinforced by the extreme slowness of the planets' movements. A complete revolution of Uranus' loops through the zodiac (84 years) requires longer than an average human life-span; for Neptune the metamorphosis takes almost double the time.

The more 'individual' motions of the two planets, however, stand in strong contrast to this impression. Although both planets are relative giants in comparison with the Earth (both are estimated to have a diameter three and a half times that of the Earth), they both appear to rotate more quickly around their own axes than the latter (Uranus $10^h 49^m$, Neptune c. 18^h). This is manifested in a considerable flattening at the poles of the two planets. With the exception of Nereid, Neptune's outer moon, all the moons revolve rapidly around their planet. Triton, the inner moon of Neptune, which is estimated to be somewhat larger than our own Moon, encircles the planet in just under six days. Of the five moons of Uranus, the slowest and outermost, Oberon, requires only $13\frac{1}{2}$ days to revolve around the planet, and Pluto's Charon remains over the same part of Pluto by revolving at the same speed as Pluto's rotation.

These autonomous, or 'individual' motions of the planets appear to be completely independent of their relationship to the Sun. It is thus characteristic that the orbital irregularities first of Uranus and then of Neptune made it possible to calculate the orbits and periods of revolution around the Sun of the two suspected, but as yet undiscovered outermost planets, though their own rotation, or the possible existence of moons could, of course, not be predicted. One could, in this sense, speak of the 'subordinate' and the 'autonomous' motions of a planet. In the case of Uranus and Neptune the subordinate motion, that is, the revolution around the Sun, possesses an outspoken Saturnine character, manifested in their extremely slow and regular movements. Their 'autonomous' motions, however, such as their own rotation and the movements of their moons, are decidedly 'Mercurial' in their rapid sequence and strong irregularities.

Seen in this light, Pluto, the outermost wanderer of our solar system, occupies an exceptional position. Its strongly eccentric and inclined orbit does not conform to the regular patterns of Uranus and Neptune. In size and appearance, moreover, it bears a much stronger resemblance to the moons of the latter planets than to the planets themselves. This has led some scientists to doubt the purely planetary character of Pluto, and the suggestion has been put forward that it may in fact be a 'liberated moon' of Neptune which broke away but did not escape from the solar system altogether.

28. The planetoids

At the beginning of the nineteenth century a small group of luminaries was discovered, which were found to display planetary motions. They were called *asteroids* (starlike) from their appearance, or *planetoids* (planet-like) from their movement. In technical works they are often called *minor planets*. The discoveries of these luminaries have increased continually and rapidly since 1845 — particularly after the introduction of celestial photography — so that a considerable swarm is now catalogued. At the end of the last century the approximate orbits of 463 planetoids were known. By 1970 roughly 7 000 such small objects had already been observed; approximate orbits had been calculated for about 1 800. Owing to the vast number of these planetoids the orbits are no longer calculated unless they are suspected to be interesting or unusual.

The first planetoid, named by its discoverer *Ceres*, was found by chance on New Year's Day, 1801. The Italian astronomer G. Piazzi was engaged in the observatory in Palermo in telescopic observations of fixed stars for the compilation of a star-catalogue. In the constellation of the Bull he noticed a star of the eighth magnitude, whose position was not registered, and which exhibited a distinct displacement on the following nights. Due to illness, however, he lost the planetoid, and only through the calculations of the twenty-four-year-old mathematician Gauss, using a newly-developed method, was it possible to find Ceres again before the end of the year. It was found that Ceres wanders between the orbits of Mars and Jupiter. Its sidereal revolution, accordingly, lies between those of the two planets. The sidereal revolution of Mars requires almost two years; that of Ceres requires four and a half, more than twice as long; while Jupiter takes roughly twelve years.

In March of the following year, 1802, Olbers in Bremen, while observing the new planet Ceres, discovered in its immediate neighbourhood a further planetoid of the seventh magnitude, which received the name of *Pallas*. In 1804 and 1807 followed the discovery of two more planetoids, which were named *Juno* and *Vesta*.

The discovery of further minor planets was interrupted until 1845, when the task of drawing star-maps of the zone around the ecliptic, containing all stars down to the ninth and, in a narrower zone, to the twelfth magnitudes, was systematically undertaken. Since this time not a year has passed without the discovery of one or several planetoids. Many have been lost again, as it was not always possible to lay aside other work in order to plot and calculate their orbits.

The brightest asteroid is Vesta, discovered in 1807, also by Olbers. At its maximum magnitude of 6 it can just be seen with the unaided eye. Only about a dozen of the minor planets attain a magnitude between 6 and 9. The great majority are extremely faint, even in the telescope.

Many of the minor planets exhibit strong variations in brightness over a period of several hours. This would seem to suggest that they are of irregular oblong shape, turning first a broader, then a narrower side towards the Earth. Their periods are normally between 4 and 17 hours, during which time the fluctuations in brightness can be considerable. Eros, for example, varies by 1.5 magnitudes in $2^h 38^m$, indicating a rotation in double this time, or $5^h 16^m$, as it twice becomes broad (and brighter) and twice narrow (and dimmer). In the case of Eros this has been confirmed by angular measurement of the disc.

The most striking common feature of the planetoids lies in the fact that their orbits are, with certain exceptions, confined to a specific region between those of Mars and Jupiter. Their heliocentric paths vary from nearly circular (for instance Adalberta) to strongly eccentric ellipses (for instance Icarus, with an eccentricity of 0.83). These planetoids often pass alternately from the realm of the inferior to that of the superior planets. Hidalgo, for instance, is beyond the orbit of Saturn at its aphelion, and yet much closer to the Sun than Mercury at its perihelion.

The minor planets differ considerably in their orbital inclinations. These can be very nearly 0°, or can climb to values as high as 34° for Pallas or 43° for Hidalgo. However, their average inclination is just under 10°. The extreme variety of phenomena offered by the minor planets and their movements has provided much useful material in the study of celestial mechanics.

Of special interest are two small groups of planetoids known as the Trojans and the Greeks. The two swarms are held captive on either side of the great planet Jupiter, with which they share their orbit. In 1772 Lagrange had demonstrated the theoretical possibility that two or more celestial bodies could share the same orbit around the Sun. Two such bodies would, according to his construction, form an equilateral triangle with the Sun; three would form two adjacent such triangles. This demonstration was long regarded as nothing more than a mathematical curiosity. At the beginning of the twentieth century, however, two groups of minor planets were in fact discovered which are distributed around these 'libration points', as they are called, 60° to either side of Jupiter. The Greek planetoids, which precede the planet in its orbit, are named, with one exception, after Greek heroes of the *Iliad*, such as Agamemnon, Achilles, Nestor and Ajax. The Trojan planetoids which follow Jupiter include Priam, Anchises, Aeneas and Troilus. These two groups of planetoids describe complex curves around the ideal libration points, but never entirely leave their immediate neighbourhood. This 'swarming' movement can be regarded as a particularly vivid illustration of the constant flux and variability which underlie all seemingly mechanical movements of the celestial bodies. The two groups of 'prisoner planets', whose movements are otherwise entirely subordinated to those of Jupiter, swarm restlessly within a small space, as though to compensate for their loss of freedom.

Although it may not be so obvious as for the two groups of Greek and Trojan planetoids, the minor planets seem, as a whole, to stand under the influence of Jupiter. This comes most clearly to expression in a number of gaps, within which no planetoids are known to circle, at distances which are an exact fraction of Jupiter's, such as a half or a third.

The planetoids, therefore, despite the exceptional variety in their orbital elements, do not really unfold an individual character in their movements. On the whole it is difficult to avert the impression of a huge expanse of cosmic debris between the orbits of Mars and Jupiter.

29. Three methods of defining a conjunction

A conjunction, the meeting of two planets in their course through the zodiac, can take place in the most manifold ways. It can be of short or of long duration. It can occur by one planet overtaking another; it can, however, also take place when two planets pass each other in opposite directions.

Owing to the different inclinations of the orbits of the Moon and the planets to the ecliptic, most conjunctions take place with a certain north-south separation. A direct occultation of one planet by another is extremely rare. Central conjunctions with the Sun or the Moon are considerably more frequent. An example is the conjunction of Mars with the Sun, which took place on November 14/15, 1944 (Figure 141). Mars, which spatially was far beyond the Sun, was overtaken and occulted by the latter in the constellation of the Scales. Mars stood at the time only 4' south of the ecliptic. Figure 141 shows in natural proportion the relative position of Sun and planet at the beginning, middle and end of the occultation. It can be seen from the figure that the conjunction lasted roughly one and a half days, from November 14, 1.00 to November 15, 12.00. The entry in a star calendar or ephemeris indicates only the mid-point of the conjunction.

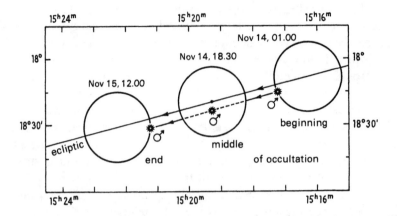

Figure 141. The occultation of Mars by the Sun at the conjunction of 1944 Nov 14/15.

The vertical separation of two planets at the moment of conjunction differs from instance to instance. In extreme cases, as the conjunction of Mercury with the Moon, it can be as much as 11°. The greater the separation, the less striking is the impression of the conjunction; and at the same time, the stronger are the differences which emerge in the mathematical calculation of the aspect.

From a spherical point of view, the exact determination of the moment of a planetary conjunction is by no means so simple as it might seem at first glance. Three distinct definitions are possible for this moment, depending on the co-ordinate system which we choose. The passage can be calculated either in right ascension (AR) *(equatorial conjunction)*, or in ecliptic longitude (λ) *(ecliptical conjunction)*; or we can choose the moment of *closest conjunction*, when the minimum spherical distance between the two planets is attained.

In order to illustrate the distinction between the different types of conjunctions,

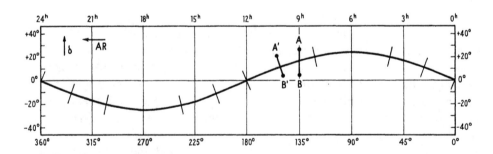

Figure 142. The ecliptic (————) in the grid of right ascension (AR) and declination (δ). The celestial equator is the central horizontal line. AB is an equatorial conjunction (on the same right ascension), A'B' is an ecliptical conjunction (on the same longitude).

Figure 142 shows the relative positions of the right ascension meridians and the ecliptic meridians (short marks along the ecliptic) for the equatorial region of the sky. The former constitute a grid based on the celestial equator. These meridians stand at right angles to the equator and meet in a point at the north and south celestial poles. The ecliptic meridians radiate from the ecliptic poles; they are perpendicular to the ecliptic, which appears in Figure 142 as a wave. An equatorial conjunction (right ascension) is therefore seen with reference to the grid of the

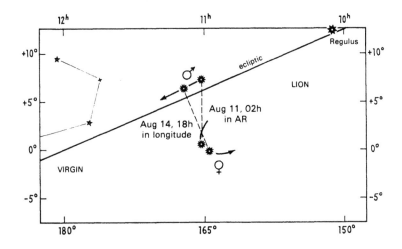

Figure 143. The conjunction of Mars and Venus in right ascension, 1959 Aug 11, and in longitude on Aug 14 (hours in UT).

celestial equator. At the moment of passage, two planets A and B stand on the same north-south line, and culminate simultaneously along a single geographical line of longitude. An ecliptical conjunction (in longitude), $A'B'$, is seen with reference to the ecliptic co-ordinates.

The greatest divergences between the two types of conjunction arise near the vernal and autumnal equinoxes. At the solstices the right ascension and ecliptic grids coincide, and the distinction between the two conjunctions vanishes.

Using three examples we shall show how considerable the deviation can be between an equatorial conjunction (right ascension) and an ecliptical conjunction (in longitude).

Figure 143 shows the mutual position of Mars and Venus in mid-August 1959 in the constellation of the Virgin. Venus is nearly stationary; Mars passes it at about 7° to the north; while Mars is about 1° north of the ecliptic. Venus is roughly 5¾° south. The equatorial conjunction takes place on August 11 at 2.00; the ecliptical conjunction follows three and a half days later on August 14 at 18.00.

Figure 144 illustrates the relative movements of Venus and Mercury in April

203

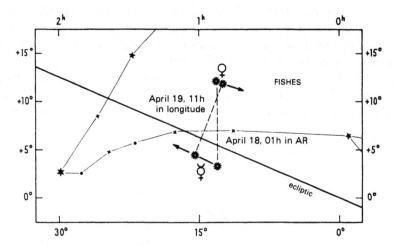

Figure 144. The two types of conjunction at the meeting of Venus and Mercury on 1961 April 18/19.

1961, in the vicinity of the vernal equinox. Venus is retrograde and stands 6° north of the ecliptic, while Mercury is 2° south of the Sun's path, so that the large separation of 8° occurs. The equatorial conjunction takes place on April 18 at 1.00, the ecliptical conjunction on April 19 at 11.00. The difference is thus nearly one and a half days.

Figure 145 shows the aspect of Venus and Jupiter at the end of February, 1964. Jupiter is overtaken in the constellation of the Fishes by Venus; both planets have direct motion. Venus stands roughly 0°.5 to the north of the ecliptic and Jupiter about 1° to the south. They are therefore separated by 1°.5. Although these differences are by no means so large as those in the previous examples, the two conjunctions are nonetheless separated by 16 hours. The ecliptical conjunction takes place on February 27 at 16.00; the equatorial conjunction on February 28, at 8.00.

Because astronomers generally use the equatorial co-ordinates in their work the conjunctions are usually calculated equatorially. However, in astrological ephemerides (such as *Raphaels*) the ecliptical conjunction is used, as the astrologer uses only ecliptical co-ordinates (see Chapter 4).

The difference between the conjunctions is smaller when the distance between the planets is less, or if the planets move more quickly (thereby passing through both conjunctions faster).

It remains for us to consider the deviations of the ecliptical and equatorial conjunctions from the *closest conjunction*. In general, the ecliptical conjunctions coincide fairly closely (much more so than the equatorial conjunctions) with the closest conjunctions. When the planets stand very close together, the differences vanish entirely.

204

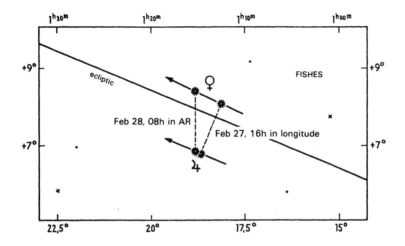

Figure 145. The conjunctions of Jupiter and Venus on 1964 Feb 27/28.

Striking differences between the closest conjunction and the ecliptical conjunction arise frequently for eclipses of the Sun (especially when they are partial). The moment of maximum occultation is the moment of closest conjunction of Sun and Moon. This can be as much as 20 minutes before or after the New Moon, as calculated for star-calendars and ephemerides on the basis of ecliptic co-ordinates.

For conjunctions of Mercury with the Sun we must take into account the movement of the planet's orbit with respect to the Earth, as illustrated in Figure 115. A representation of the deviations between the closest and the ecliptical conjunctions for Mercury's inferior conjunctions with the Sun is given in Figure 146. (A similar, but somewhat modified picture is applicable to the superior conjunctions.) Each curve represents the portion of Mercury's orbit which is nearer the Earth at a given time. In Figure 146 twelve characteristic positions are shown, representing the orbit's change of orientation during the course of a year. A horizontal line (not drawn) through the Sun in the centre of the figure would correspond to the ecliptic. The vertical dotted line indicates the positions of the ecliptical conjunctions. Because of the changing inclination of Mercury's orbit, the closest conjunction usually deviates either to the west (right) or to the east (left). Along the lemniscate lie all the points of closest conjunction of Mercury and the Sun.

At an inferior conjunction the planet's deviation from the Sun can attain a maximum of ± 4°.2 (the height of the diagram); at superior conjunctions it is confined to within ± 2°. The difference results from the spatial perspective in which Mercury's movements appear as its distance from the Earth changes.

The same is true in principle for Venus. Its closest conjunctions with the Sun are also distributed along a slender lemniscate, in whose centre the Sun stands,

205

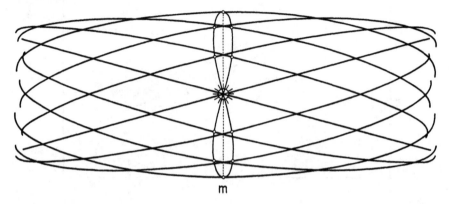

m

Figure 146. A comprehensive picture of the deviations between the true and ecliptical conjunctions of Mercury with the Sun.

and the length of which is dependent on the planet's greatest possible deviation from the ecliptic. For the inferior Venus conjunctions, therefore, the height of the lemniscate is 14°.75, while the lemniscate belonging to the superior conjunction is foreshortened in perspective to only 2°.8.

The conjunctions of the other planets with the Sun and with each other can be portrayed in an exactly corresponding manner.

It is altogether an interesting and important feature of the closest conjunctions that they are governed by principles whose formal expression is to be sought in characteristic lemniscates. In an earlier chapter we have seen how a similar principle underlies the fluctuating spatial relationship between the Earth and the Sun over the course of a year. That this is not the case for the ecliptical and equatorial conjunctions is evident when we consider that the former are determined exclusively on the basis of the Sun's movement, and the latter on the basis of the Earth's movement. Because the closest conjunctions are not related to a single system of co-ordinates, but arise as a faithful reflection of the mutual motions of Earth, Sun and planets, they may be regarded as an especially characteristic type of conjunction, whose celestial signature most effectively reveals the harmonious nature of the underlying movements.

The equatorial conjunction reflects the moment when the two luminaries culminate simultaneously. It is thus related to the time of day, and to a certain geographical longitude, to a place on the Earth.

The ecliptical conjunction relates the luminaries to the Sun, in that it uses the Sun's annual path as its reference. The closest conjunction is independent of all reference systems and speaks most readily to direct observation.

30. The Great Conjunction

The conjunctions of two planets are repeated in more or less pronounced rhythms of shorter or longer duration. In ancient times particular heed was paid to the so-called Great Conjunction of Saturn and Jupiter.

Owing to their long periods of revolution through the zodiac, the meetings of Saturn and Jupiter take place at greater intervals of time. Such conjunctions are repeated periodically roughly every twenty years.

Two Saturn revolutions of 30 years each very nearly correspond to five Jupiter revolutions of 12 years each. This means that in 60 years Jupiter makes 5 revolutions, and Saturn 2. If they begin at a conjunction, Jupiter overtakes Saturn three times (5–2), in other words, every 20 years. Expressed more exactly, the mean period lasts 19.86 years.

Moreover, the relative movements of the two planets brings about a displacement from one conjunction to the next of two-thirds of the total circumference of the zodiac, that is, by 240° or 8 constellations (Figure 147). Only after the passage of sixty years (3 × 20) does a conjunction take place in the same region as at the beginning. In this manner, the conjunctions are confined over a longer span of time to three constellations, forming a trigon in the zodiac. At present they occur in the constellations Ram, Archer and Virgin. The conjunctions between 1800 and 2000 lie as follows:

Year	Constellation
1802	Lion
1821/22	Fishes
1842	Archer
1861	Lion
1881	Ram
1901	Archer
1921	Virgin
1940/41	Ram
1961	Archer
1980/81	Virgin
2001	Ram

As the separation between two successive conjunctions is not exactly equal to 240°, but is slightly more (242°.7) the resulting conjunction-trigon is an open

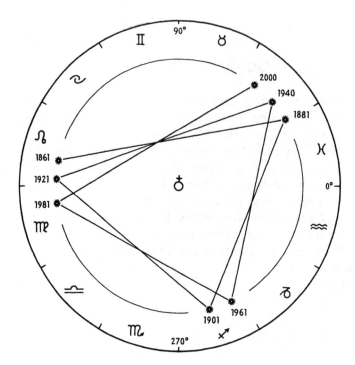

Figure 147. The trigon of the Saturn-Jupiter conjunctions and its displacement in the zodiac.

figure which slowly rotates in forward (anticlockwise) direction through the zodiac (Figure 147). The conjunctions of Saturn and Jupiter do not, therefore, return every 60 years to exactly the same position in the zodiac, but are displaced by 8° to the east. After 220 years the displacement has reached 30°, and after 4 times this period, or about 880 years, it has increased to 120°, when the points of conjunction coincide once more with their three starting-points.

The rotation of the positions we have now considered takes place relative to the fixed stars, or the constellations of the zodiac; the sidereal revolutions of the planets are used as a basis of calculation. We can also consider the positions relative to the vernal equinox and thus to the regular signs of the zodiac (see Chapter 5); for this the tropical periods of the planets are used.

Because of the slow retrograde movement of the equinox through the zodiac, the Saturn-Jupiter conjunctions return to it with a slightly shorter period than to a fixed point in the zodiac. Relative to the equinox the conjunctions advance by 9° in 60 years, and by 27° in 180 years. After 200 years, or ten conjunctions, the trigon has advanced by nearly one complete sign of 30°. For the next 200 years, therefore, the conjunctions have passed into three new signs, which constitute a new trigon. After 800 years they have passed through all four trigons in the zodiac. Forty conjunctions take place during the whole of this period. Its exact length is

30. THE GREAT CONJUNCTION

794.4 years. At its conclusion, a meeting of Saturn and Jupiter will have advanced through all twelve signs of the zodiac and have returned to the vernal equinox.

The individual conjunctions of Saturn and Jupiter take on manifold forms. If they take place soon before or after the conjunction of the two planets with the Sun, as was the case for example in 1961, only a single meeting takes place, and is either invisible, or restricted to a brief visibility in the morning or evening twilight. The most impressive conjunctions are those which occur during the planets' loop-formation near opposition to the Sun. In such cases, which do not recur at regular intervals, a triple meeting of the two planets takes place. Jupiter, in its more broadly sweeping loop-movement, passes by Saturn three times. The triple Saturn-Jupiter conjunctions from December 1980 to July 1981 bore this pronounced character.

The tropical revolution of the conjunctions, embracing almost eight centuries and falling into four natural subdivisions of two centuries each, was known in ancient star-lore, as well as in medieval times, as the rhythm of the Great Conjunction. An exceptional significance was attached in particular to the passage of the positions of conjunction through the vernal equinox. This point in time was reckoned as the beginning of a new eight-hundred-year conjunction-period. The last such passage took place in 1593, at the time of Kepler. Kepler himself devoted much attention to the rhythms of the meetings between Jupiter and Saturn. In the foreword to his early work *Mysterium Cosmographicum* (1596) he describes in detail how the construction of the trigon figures of the Great Conjunction led him to the idea which became the most fundamental feature of his astronomical research during the whole of his life: proceeding from the regular geometrical figures and the Platonic solids, to discover the proportions and harmonic relationships between the sizes and separations of the different planetary orbits.

Geocentric and heliocentric world-views

Epilogue by Suso Vetter

When we attempt to gain a conceptual understanding of the nature of the celestial phenomena we soon find ourselves facing a dilemma which is characteristic for the age in which we live. In observing the stars day after day and year after year, we have the immediate experience of standing on the motionless Earth and watching the stars overhead in their movements. If, however, we open an astronomy textbook, we read that the observed movements are only 'apparent'; that the Sun stands at the centre of the planetary movements, and that the Earth also revolves about this central star. This heliocentric (Sun-centred) conception, which contradicts at first the observed phenomena, can only be grasped intellectually, not empirically.

The heliocentric world-view is, moreover, by no means so obvious as is generally assumed. For our whole life unfolds naturally under the geocentric conditions. The astronomical calendars and ephemerides, as well as all aspects of practical astronomy, form no exception to this.

We can see from the foregoing how two separate complexes of questions arise out of the observation of the celestial phenomena, which must be distinguished clearly at first, in order then to be reconciled.

The first centres around the inquiry into the observed phenomena, movements and rhythms of the luminaries in relation to the Earth. And in this relationship lies a reality, to which reference has already been made in the introduction. This aspect alone underlies the phenomena dealt with in the present volume.

The other complex of questions concerns the 'true' movements of the celestial bodies. It had been the intention of Joachim Schultz to treat this problem in a later volume with reference to the various conceptions of the structure of the cosmos as they have emerged during the course of history. A few basic remarks on this theme may here be allowed.

The geocentric world-view of ancient times was founded in the prevailing consciousness of man in his whole relationship to Earth and Cosmos. Plato's maxim that all celestial phenomena could be reduced to regular circular movements led to the postulation of a complicated system of mutually overlapping circular rotations (deferents and epicycles).

A knowledge of the heliocentric point of view, which was put forward in Greek times by Aristarchus of Samos, was consciously suppressed. Copernicus read about this world-view in ancient sources, and hoped that its confirmation would lead to an essential simplification of the extremely complicated geocentric system of the planetary movements. Kepler was also acquainted with the ancient accounts, and was stimulated by Copernicus to further investigations. With the help of Tycho Brahe's decades of observations, Kepler was able to calculate the planetary orbits as ellipses with definite properties, known as the orbital elements.

In this system the revolution of a planet around the Sun through the constellations of the zodiac is defined as its sidereal revolution. An important property of these elliptical orbits is their inclination to the plane of the ecliptic, which is here defined as the Earth's orbit. In the chapter on the three outermost planets Uranus, Neptune and Pluto, reference is made to these properties.

Appendix

1. The Stars

Table 1.1 Symbols and abbreviations

Signs (or constellations) of the zodiac

♈ Aries	the Ram	♎ Libra	the Scales or Balance
♉ Taurus	the Bull	♏ Scorpius	the Scorpion
♊ Gemini	the Twins	♐ Sagittarius	the Archer
♋ Cancer	the Crab	♑ Capricornus	the Mountain-Goat or Sea-Goat
♌ Leo	the Lion	♒ Aquarius	the Water-Bearer
♍ Virgo	the Virgin	♓ Pisces	the Fishes

The planets

☉ Sun	☿ Mercury	♅ Uranus
☾ Moon	♀ Venus	♆ Neptune
♁ or ⊕ Earth	♂ Mars	♇ Pluto
	♃ Jupiter	
	♄ Saturn	

Ag   apogee (farthest from the Earth)

Pg   perigee (closest to the Earth)

Ah   ahelion (farthest from the Sun)

Ph   perihelion (closest to the Sun)

The Moon

🌑 New Moon

🌓 First Quarter   ☊ ascending node

🌕 Full Moon   ☋ descending node

🌗 Last Quarter

Aspects

☌ Conjunction (0°)	□ Quadrature (90°)	✳ Sextile (60°)
☍ Opposition (180°)	△ Trigon (120°)	

Magnitudes

Stars (and planets) are grouped into classes of magnitude according to brightness. Each class is 2.512 times as bright as the preceding. The higher the magnitude (mag) the fainter the star appears. Stars up to about mag 6.0 are visible to the naked eye.

APPENDIX

Table 1.2 Co-ordinate systems.

a. *Horizontal system*
   Azimuth (A): horizontal plane from south* westward (0° to 360°)
   Altitude (h): vertically from horizon, +ive upwards† (−90° to +90°)

b. *Equatorial system*
   (i) Fixed, hour angle system
   Hour angle (τ): equatorial plane from meridian* westward (0ʰ to 24ʰ)
   Declination (δ): perpendicular to equator, +ive north (−90° to +90°)
   (ii) Moving, Right Ascension system
   Right Ascension (α): Equatorial plane from vernal equinox eastward (0ʰ to 24ʰ)
   Declination (δ): as above

c. *Ecliptical system*
   Longitude (λ): ecliptical plane from vernal equinox eastward (0° to 360°)
   Latitude (β): perpendicular to ecliptic, +ive north (−90° to +90°)

* In southern hemisphere measure from north meridian westward
† In southern hemisphere +ive *down*ward

φ = terrestrial latitude (+ive north)
ε = angle of ecliptic to equator = 23°46205 = 23° 27′ 43″4 (1975.0)
S = Local Sidereal Time (LST)

Table 1.3 Conversion of co-ordinate systems

From horizontal to fixed equatorial
$$\sin \delta = \sin h \times \sin \varphi - \cos h \times \cos A \times \cos \varphi$$

$$\sin \tau = \frac{\cos h \times \sin A}{\cos \delta}$$

From fixed equatorial to horizontal
$$\sin h = \sin \delta \times \sin \varphi - \cos \delta \times \cos \tau \times \cos \varphi$$
$$\sin A = \frac{\cos \delta \times \sin \tau}{\cos h}$$

From fixed to moving equatorial
$\alpha = S - \tau$ (In southern hemisphere $\alpha = S - \tau - 12^h$)
$\delta = \delta$

From moving to fixed equatorial
$\tau = S - \alpha$ (In southern hemisphere $\tau = S - \alpha - 12^h$)
$\delta = \delta$

From moving equatorial to ecliptical
$$\sin \beta = -\cos \delta \times \sin \alpha \times \sin \varepsilon + \sin \delta \times \cos \varepsilon$$
$$\cos \lambda = \frac{\cos \delta \times \cos \alpha}{\cos \beta}$$

From ecliptical to moving equatorial
$$\sin \delta = \cos \beta \times \sin \lambda \times \sin \varepsilon + \sin \beta \times \cos \varepsilon$$
$$\cos \alpha = \frac{\cos \beta \times \cos \lambda}{\cos \delta}$$

Table 1.4 Apparent magnitude and position of stars

Following is a list of all stars down to mag 1.5 and a few others, showing the name, magnitude and position for 1975.0 in Right Ascension and declination.

Constellation	Name	Mag	AR h m s	δ ° '
α Cassiopeia	Shedar	2.3	0 39 05	+ 56 24
α Eridanus	Achernar	0.6	1 36 47	− 57 22
α Fishes	Al Rischa	4.3	2 00 45	+ 2 39
α Ram	Hamal	2.2	2 05 46	+ 23 21
α Lesser Bear	Polaris	2.1	2 07 26	+ 89 09
α Bull	Aldebaran	1.1	4 34 29	+ 16 28
β Orion	Rigel	0.3	5 13 20	− 8 14
α Charioteer	Capella	0.2	5 14 50	+ 45 58
γ Orion	Bellatrix	1.7	5 23 47	+ 6 20
β Bull	El Nath	1.8	5 24 43	+ 28 35
ε Orion	Alnilam	1.7	5 34 57	− 1 13
α Orion	Betelgeuze	0 − 1	5 53 49	+ 7 24
α Keel of Ship	Canopus	−0.9	6 23 24	− 52 41
α Greater Dog	Sirius	−1.6	6 44 03	− 16 41
α Twins	Castor	1.6	7 33 00	+ 31 57
α Lesser Dog	Procyon	0.5	7 38 00	+ 5 17
β Twins	Pollux	1.2	7 43 47	+ 28 05
α Crab	Acubens	4.3	8 57 07	+ 11 57
α Lion	Regulus	1.3	10 07 02	+ 12 05
α Greater Bear	Dubhe	1.9	11 02 12	+ 61 53
β Lion	Denebola	2.2	11 47 47	+ 14 43
α Cross	Acrux	1.6	12 25 12	− 62 58
β Cross		1.5	12 46 15	− 59 33
α Virgin	Spica	1.2	13 23 52	− 11 02
β Centaur	Agena	0.9	14 02 03	− 60 15
α Bootes	Arcturus	0.2	14 14 31	+ 19 19
α Centaur	Rigil Centauris	0.1	14 37 53	− 60 44
β Scales	Kiffa Borealis	2.7	15 15 40	− 9 18
α Scorpion	Antares	1.2	16 27 52	− 26 23
α Lyre	Vega	0.1	18 36 06	+ 38 46
α Eagle	Altair	0.9	19 49 34	+ 8 48
α Swan	Deneb	1.3	20 40 35	+ 45 11
δ Goat	Deneb Algedi	3.0	21 45 40	− 16 14
α Water-Bearer	Sadalmelik	3.2	22 04 30	− 0 27
α Southern Fish	Fomalhaut	1.3	22 56 16	− 29 45

APPENDIX

2. The Sun

Table 2.1 Distance and size of the Sun

	AU	km	miles
Mean distance Sun – Earth	1.0000	149 600 000	93 000 000
Greatest distance	1.0167	152 000 000	94 450 000
Least distance	0.9833	147 000 000	91 350 000

Apparent diameter at mean distance 31′ 59″26
(at greatest distance 31′ 28″9; at least distance 32′ 33″1)
Estimated true diameter 1 392 000 km, 864 000 miles
Inclination of solar equator to ecliptic 7° 15′

Table 2.2 Daily and annual rhythms of the Sun

Mean solar day	24^h		
Sidereal day	23^{h}934 471 93	= 23^h 56^m 4^{s}09895	
Difference	0^{h}065 528 07	= 3^m 55^{s}90105	

	Length	*Difference from tropical*
Tropical year	365^{d}242 203 40	
(equinox to equinox)	365^d 5^h 48^m 46^{s}374	
Sidereal year	365^{d}256 360 50	0^{d}014 157 10
(star to star)	365^d 6^h 9^m 9^{s}547	20^m 23^{s}173
Apsidal year	365^{d}259 643 62	0^{d}017 440 22
(perigee to perigee)	365^d 6^h 13^m 53^{s}208	25^m 6^{s}835

Mean daily movement of Sun on ecliptic	59′ 8″3
Smallest daily movement (in July)	57′ 10″4
Greatest daily movement (in January)	61′ 9″4

Length of summer half-year	186½ days
Length of winter half-year	178¾ days
Difference	7¾ days

Table 2.3 The equation of time

Following values may be taken as a guide for any year. Exact values for a particular year may be found in an ephemeris.

Jan 1	− 3^m 26^s	11	− 1^m 13^s	Jul 10	− 5^m 3^s	18	+14^m 29^s
11	− 7^m 51^s	21	+ 1^m 12^s	20	− 6^m 8^s	28	+16^m 5^s
21	−11^m 19^s	May 1	+ 2^m 55^s	30	− 6^m 17^s	Nov 7	+16^m 15^s
31	−13^m 31^s	11	+ 3^m 45^s	Aug 9	− 5^m 28^s	17	+15^m 4^s
Feb 10	−14^m 23^s	21	+ 3^m 38^s	19	− 3^m 40^s	27	+12^m 29^s
20	−13^m 57^s	31	+ 2^m 38^s	29	− 1^m 3^s	Dec 7	+ 8^m 42^s
Mar 2	−12^m 24^s	Jun 10	+ 0^m 56^s	Sep 8	+ 2^m 8^s	17	+ 4^m 5^s
12	−10^m 2^s	20	− 1^m 11^s	18	+ 5^m 38^s	27	− 0^m 53^s
22	− 7^m 10^s	30	− 3^m 17^s	28	+ 9^m 7^s	Jan 6	− 5^m 38^s
Apr 1	− 4^m 6^s			Oct 8	+12^m 14^s		

Table 2.4   The Sun in the zodiac

Entry of the Sun into the constellations and signs

			Constellation		Sign		
Apr	18	28°	Ram	Aries	0°	Mar	21
May	14	53°	Bull	Taurus	30°	Apr	21
Jun	20	89°	Twins	Gemini	60°	May	22
Jul	20	117°	Crab	Cancer	90°	Jun	22
Aug	11	138°	Lion	Leo	120°	Jul	23
Sep	16	173°	Virgin	Virgo	150°	Aug	24
Nov	1	218°	Scales	Libra	180°	Sep	24
Nov	20	237°	Scorpion	Scorpius	210°	Oct	24
Dec	19	267°	Archer	Sagittarius	240°	Nov	23
Jan	20	299°	Goat	Capricornus	270°	Dec	22
Feb	15	326°	Water-Bearer	Aquarius	300°	Jan	21
Mar	13	352°	Fishes	Pisces	330°	Feb	19

Table 2.5   Precession and apsidal movement

Annual precession	50.″273	(for 1975.0) westwards	
In 72 years		1°	
in 72 × 30	=	2160 years	30°
in 72 × 30 × 12	=	25 920 years	360° = Platonic year

Annual apsidal movement 11.″8 eastward
Period of rotation about 110 000 years

Table 2.6   The vernal equinox in the zodiac

Constellation	Date of entry	Duration
Virgin	BC 13800	3 300 years
Lion	BC 10500	2 500
Crab	BC 8000	1 500
Twins	BC 6500	2 100
Bull	BC 4400	2 600
Ram	BC 1800	1 700
Fishes	BC 100	2 700
Waterman	AD 2600	1 800
Goat	AD 4400	2 100
Archer	AD 6500	2 100
Scorpion	AD 8600	2 300
Scales	AD 10900	1 300
Virgin	AD 12200	

APPENDIX

3 The Moon

Table 3.1 Distance and size of the Moon

	from poles	from equator
Apparent diameter at mean distance	31' 05"	31' 37"
Apparent size at (mean) greatest distance	33' 31"	34' 08"
Apparent size at (mean) least distance	29' 23"	29' 51"

True diameter	3476 km	2160 miles	0.272 Earth diameter	
Greatest magnitude	−12ᵐ5	Albedo 0.07		

Eccentricity	$e = 0.05490$	($\pm$ 0.0117)

	km	miles	A U	Earth diameters	Moon
Mean distance Earth – Moon	384 400	239 000	0.00257	30.1	111
Mean greatest distance	405 500	252 100	0.00271	31.8	117
Mean least distance	363 300	225 900	0.00243	28.5	105

Mean inclination to ecliptic	5° 09'
Mean inclination to lunar equator	6° 41'

Table 3.2 Monthly lunar rhythms

	d	d h m s
Synodic (New Moon to New Moon)	29.530 589	29 12 44 02.9
Tropical (equinox to equinox)	27.321 582	27 7 43 04.7
Sidereal (fixed star to fixed star)	27.321 661	27 7 43 11.5
Nodal (Moon's node to Moon's node)	27.212 220	27 5 05 35.8
Apsidal (perigee to perigee)	27.554 551	27 13 18 33.2

	synodic	sidereal
Mean daily movement	12° 11' 27"	13° 10' 35"
Mean hourly movement	30' 29"	32' 56"

Table 3.3 Other lunar rhythms

1 lunar year: 12 synodic months	354ᵈ36705	= 354ᵈ 8ʰ 48ᵐ 36ˢ
1 tropical solar year	365ᵈ24220	= 365ᵈ 5ʰ 48ᵐ 46ˢ
difference	10ᵈ87515	≃ 11ᵈ
Metonic cycle: 235 synodic months	6939ᵈ68819	= 6939ᵈ 16ʰ 31ᵐ
19 tropical years	6939ᵈ60208	= 6939ᵈ 14ʰ 27ᵐ
difference	0ᵈ08611	= 2ʰ 4ᵐ
1 revolution of nodes 18.5997 years	6793ᵈ39	= 18ᵃ 218ᵈ 21ʰ 22ᵐ
Regression of nodes in 1 year	19° 21'.5	
Regression of nodes in 1 month about	1°5	
1 revolution of apsides	8.8508 years	= 3232ᵈ6
Progression in one year (average)	40° 41'	
1 Saros period: 223 synodic rhythms	6585ᵈ321	
242 nodal rhythms	6585ᵈ357	
239 apsidal rhythms	6585ᵈ537	

(Further eclipse rhythms in G. van den Bergh, *Periodicity*)

4. The planets

Table 4.1 Distance of the planets

Planet	Distance from Earth (AU)			Distance from Sun (AU)			Eccentricity
	Mean	Least	Greatest	Mean	Least	Greatest	
Moon	0.00263	0.00238	0.00272	—	—	—	0.055
Mercury	1.00	0.53	1.47	0.39	0.31	0.47	0.2056
Venus	1.00	0.27	1.73	0.72	0.72	0.73	0.0068
Sun	1.00	0.98	1.02	—	—	—	—
Earth	—	—	—	1.00	0.98	1.02	0.0167
Mars	1.52	0.38	2.67	1.52	1.38	1.67	0.0934
Jupiter	5.20	3.95	6.45	5.2	4.95	5.45	0.0485
Saturn	9.53	8.00	11.07	9.55	9.01	10.07	0.0556
Uranus	19.18	17.29	21.07	19.20	18.29	20.07	0.0472
Neptune	30.05	28.80	31.31	30.09	29.79	30.33	0.0086
Pluto	39.5	28.6	50.3	39.5	29.6	49.3	0.253

Table 4.2 The planetary bodies

Planet	Colour	Magnitude	Albedo	Apparent diameter		Flat-tening	Inclina-tion*	Period of rotation	No of satellites
				Min.	Max.				
Earth	bluish		0.39			0.0034	23° 27'	23h 56m 4s	1
Moon	yellow-white	Full: −12	0.07	29' 22"	33' 31"	0	6° 40'	27d 7h 43m 12s	—
Mercury	yellow-white	−1.6 to +3.5	0.06	4".5	13"	0	6°	58d 16h	
Venus	brilliant white	−4.5 to −3.5	0.72	10"	1' 05"	0	7°	243d 2h	
Sun	white	−26.7	—	31' 31"	32' 36"	0	7° 10'	25d to 34d	9
Mars	reddish	−2.8 to +2.0	0.15	3".5	26"	0.005	25° 12'	24h 37m 23s	2
Jupiter	yellow-white	−2.4 to −1.2	0.70	30"	50"	0.065	3° 7'	9h 50m	13+
Saturn	pale yellow-white	−0.2 to +1.5	0.75	15"	20"	0.108	26° 45'	10h 10m	22+
Uranus	greenish	+6	0.90	3"	4"	0.030	97° 59'	10h 49m	5
Neptune	bluish	+8	0.82	2".5	2".7	0.026	29°	~18h	2
Pluto	yellow	+15	0.9?		0".3	?	118°?	6d 9h	1

*Inclination of planet's equator to planet's orbit.

220

Table 4.3 The satellites of the planets

Satellite	Mean sidereal revolution d h	Distance		Estimated diameter km
		(planet diameters)	1000 km	
Mars				
Phobos	0 7.6	1.4	9.4	20*
Deimos	1 6.3	3.5	23.5	13*
Jupiter				
V Amalthea	0 11.9	1.3	181	160*
I Io	1 18.5	3.0	422	3 620
II Europa	3 13.2	4.7	671	3 050
III Ganymede	7 3.7	7.5	1 070	5 260
IV Callisto	16 16.5	13	1 883	4 800
XIII Leda	239	78	11 094	20
VI Himalia	250 15	80	11 480	180
X Lysithea	259 5	82	11 720	40
VII Elara	259 17	82	11 740	80
XII Ananke	631 *R*	148	21 200	30
XI Carme	692 *R*	158	22 600	40
VIII Pasiphae	735 *R*	165	23 500	50
IX Sinope	758 *R*	166	23 700	40
and at least three others (very small)				
Saturn				
Mimas	0 22.6	1.5	186	390
Enceladus	1 8.9	2.0	238	500
Tethys	1 21.3	2.4	295	1 050
Dione	2 17.7	3.1	377	1 120
Rhea	4 12.4	4.4	527	1 530
Titan	15 22.7	10.1	1 222	5 150
Hyperion	21 6.6	12.3	1 483	260*
Iapetus	79 7.8	29.7	3 560	1 440
Phoebe	550 8 *R*	107.9	12 950	160
and at least thirteen others (very small)				
Uranus				
Miranda	1 10	2.6	130	400
Ariel	2 12.5	3.8	192	1 300
Umbriel	4 3.5	5.3	267	1 100
Titania	8 16.9	8.6	438	1 600
Oberon	13 11.1	11.5	586	1 600
Neptune				
Triton	5 21.1 *R*	7.2	355	3 800
Nereid	360 5	113	5 510	300
Pluto				
Charon	6 10	7	20	1 000?
R = retrograde motion	* = irregular shaped satellite			

221

Table 4.4 Revolution in the zodiac

Planet	Geocentric (with respect to Earth)		Heliocentric (with respect to the Sun)				
	One complete revolution*	Segment of 30°	Sidereal revolution	Mean daily motion	Incli-nation (1)	Position of node (2)	Position of peri-hel (3)
Moon	27^d 7^h 43^m	2½ days	—	—	—	—	—
Mercury	1 year	14 days to 9 wks	87^d.9690 ≃ 88^d	4° 5′ 32″	7°.00	48°	77°
Venus	10–13 mths	3D wks to 4 mths	224^d.7008 ≃ 225^d	1° 36′ 8″	3°.39	76°	131°
Sun	365^d 6^h 9^m	30 days	—	—	—	—	—
Earth	—	—	365^d.256 = 1^a	59′ 8″	—	—	—
Mars	almost 2 yrs	6 wks to 6 mths	686^d.9798 ≃ 1^a 322^d	31′ 27″	1°.85	49°	335°
Jupiter	12 years	1 year	4332^d.588 ≃ 11^a 315^d	4′ 59″	1°.31	100°	14°
Saturn	29 years	2½ years	10 759^d.2 ≃ 29^a 167^d	2′ 5″	2°.49	113°	92°
Uranus	84 years	7 years	84^a 8^d	42″.23	0°.77	74°	172°
Neptune	165 years	13S years	164^a 282^d	21″.53	1°.77	131°	47°
Pluto	248 years	20S years	247^a 257^d	14″.3	17°.16	110°	223°

(1) Orbital inclination to ecliptic   (2) Heliocentric ascending node in heliocentric longitude   (3) in heliocentric longitude
*Owing to loop formations the geocentric movements are subject to considerable variations, especially of Mercury, Venus and Mars.

Table 4.5 Synodic periods

Planet	Approximate value	Mean value	Variation
Moon	29½ days	29^d.530588 = 29^d 12^h 44^m 2^s.8	±7^h
Mercury	4 months	115^d.8774 ≃ 116^d	104^d – 132^d
Venus	1 year, 7 months	583^d.9205 ≃ 1^a 219^d	577^d – 592^d
Mars	2 years, 2½ months	779^d.9382 ≃ 2^a 49^d	764^d – 810^d
Jupiter	1 year, 1 month	398^d.8846 ≃ 1^a 33^d	395^d – 404^d
Saturn	1 year, ½ month	378^d.0928 ≃ 1^a 13^d	⎫
Uranus	slightly	369^d.66 ≃ 1^a 4^d	⎬ slight
Neptune	over	367^d.48 ≃ 1^a 2^d	⎭
Pluto	1 year	366^d.72 ≃ 1^a 1½^d	

Table 4.6   Forward and retrograde motion

Planet	Forward motion in one synodic period		Retrograde motion in one synodic period		Number of loops
	in degrees	in days	in degrees	in days	
Moon	387°	29½^d	—	—	—
Mercury	{ 114° (112°5 – 130°)	94^d (88^d – 102^d)	} 12°5 (±4°)	22^d (±3^d)	{ 3 in 1 year 22 in 7 years
Venus	576°	542^d ≈ 18 mths	16°	42^d	5 in 8 years
Mars	410°	706^d ≈ 25 mths	15° (4°5)	74^d (60^d – 82^d)	7 in 15 years
Jupiter	43°	279^d ≈ 9 mths	10°	120^d ≈ 4 mths	11 in 12 years
Saturn	18°	238^d ≈ 8 mths	6°5	140^d ≈ 4½ mths	28 in 29 years
Uranus	4°	218^d ≈ 7 mths	3°	152^d ≈ 5 mths	83 in 84 years
Neptune	3°	210^d ≈ 7 mths	2°5	158^d ≈ 5 mths	164 in 165 years
Pluto	1°5	200^d ≈ 7 mths	1°	165^d ≈ 5½ mths	247 in 248 years

Table 4.7   Planetary periods
Comparison of solar (tropical) years and planet's synodic and sidereal periods

			Diff. from synodic
Mercury			
		d	d
	1 year	365.24	−17.61
	3 synodic	347.63	
	4 sidereal	351.87	+ 4.24
	7 years	2 556.70	− 7.39
	22 synodic	2 549.30	
	29 sidereal	2 551.10	+ 1.80
	13 years	4 748.15	+ 2.82
	41 synodic	4 750.97	
	54 sidereal	4 750.33	− 0.65
	46 years	16 801.14	+ 1.08
	145 synodic	16 802.22	
	191 sidereal	16 802.08	− 0.14
	79 years	28 854.13	− 0.66
	249 synodic	28 853.47	
	328 sidereal	28 853.83	+ 0.36
Venus			
	8 years	2 921.94	− 2.34
	5 synodic	2 919.60	
	13 sidereal	2 921.11	+ 1.51

Mars			
	2 years	730.48	+49.45
	1 synodic	779.94	
	1 sidereal	686.98	−92.96
	15 years	5 478.63	−19.07
	7 synodic	5 459.57	
	8 sidereal	5 495.84	+36.27
	47 years	17 166.38	− 7.74
	22 synodic	17 158.64	
	25 sidereal	17 174.49	+15.85
	79 years	28 854.13	+ 3.58
	37 synodic	28 857.71	
	42 sidereal	28 853.15	− 4.56
Jupiter			
	12 years	4 382.91	+ 4.82
	11 synodic	4 387.73	
	1 sidereal	4 332.59	−55.14
	71 years	25 934.75	− 7.25
	65 synodic	25 927.50	
	6 sidereal	25 995.53	+68.03
	83 years	30 315.10	+ 0.13
	76 synodic	30 315.23	
	7 sidereal	30 328.12	+12.89
Saturn			
	29 years	10 592.0	− 5.4
	28 synodic	10 586.6	
	1 sidereal	10 759.2	+172.6
	59 years	21 549.3	+ 2.0
	57 synodic	21 551.3	
	2 sidereal	21 518.4	−32.9
Uranus			
	84 years	30 680.3	+ 0.6
	83 synodic	30 680.9	
	1 sidereal	30 688.3	+ 7.4

APPENDIX

Table 4.8   A historical note on planetary periods

A knowledge of the periodic return of celestial phenomena belonged to the most important features of ancient star-lore. Aside from the rhythms of Sun and Moon, it is documented that the late Babylonians knew of the planetary periods of 8 years for Venus; of 79 years for Mars, of 83 years for Jupiter and of 59 years for Saturn. The eight-year Venus rhythm was undoubtedly known to the early Babylonians as well.

Hipparchus (120 BC), according to the testimony of Ptolemy (*Almagest* IX), knew of the period of 59 years for Saturn; of 71 years for Jupiter, of 79 years for Mars, of 8 years for Venus and of 46 years for Mercury.

The medieval planetary tables, the *Alfonsine Tables* named after Prince Alfons of Castilia, appeared around 1250 and contained ephemerides for the individual planets for the periods of their cycles. The positions of Saturn were given for 59 years, those of Jupiter for 83 years, of Mars for 79 years, of Venus for 8 years, and of Mercury for 46 years.

These and other planetary periods were of great importance in ancient cultures for the measurement of time and for the arrangement of the calendar. In the whole of East Asia, for example, in India, Tibet, China and Japan, the twelve-year Jupiter period has been in use since the earliest times as the foundation of a twelve-year cycle. In China and Japan the individual years are still named and numbered according to this system.

In the ancient Mexican Maya culture the Venus-rhythms played an important role in the division of time in the pre-Christian millennia, alongside the rhythms of Sun and Moon.

5 The calendar

Table 5.1 Leap years

The civil calendar-year has 365 days. Every four years an extra day is added, February 29, to make 366 days (a leap year). All leap years are divisible by 4.

If a century's last year is not divisible by 400, it is not a leap year, for instance 1700, 1800, 1900, 2100 and so on (Gregorian calendar reform of 1582).

Table 5.2 Easter

Easter is the first Sunday following the first Full Moon after the vernal equinox. The earliest and latest dates of Easter are March 22 and April 25 respectively. In practice Easter is now determined by Gauss' rule:

let z be the year in question

a be the remainder after division of $z/19$

b be the remainder after division of $z/4$

and c be the remainder after division of $z/7$

For Gregorian calendar, values of x and y are as follows:

1800 to 1899 $x = 23$ $y = 4$

1900 to 2099 $x = 24$ $y = 5$.

Now find d, the remainder after division $\dfrac{19a + x}{30}$

and e, the remainder of $\dfrac{2b + 4c + 6d + y}{7}$

Easter is on   March $(22 + d + e)$

or   April $(d + e - 9)$

Write April 19 instead of April 26.

If $d = 28$ and a is greater than 10, write April 18 instead of April 25.

Table 5.3 The dates of Easter 1981 to 2010

1981 April 19	1991 March 31	2001 April 15
1982 April 11	1992 April 19	2002 March 31
1983 April 3	1993 April 11	2003 April 20
1984 April 22	1994 April 3	2004 April 11
1985 April 7	1995 April 16	2005 March 27
1986 March 30	1996 April 7	2006 April 16
1987 April 19	1997 March 30	2007 April 8
1988 April 3	1998 April 12	2008 March 23
1989 March 26	1999 April 4	2009 April 12
1990 April 15	2000 April 23	2010 April 4

226

APPENDIX

6 Ephemeris 1981–2010

Table 6.1 Solar eclipses

Date	Time	Type	Central shadow belt passes through:
1981 Feb 4	22.08	annular	Great Australian Bight, New Zealand (Stewart Is.), SE Pacific
Jul 31	3.46	total	E Black Sea, Siberia, Japan (Sakhalin), Hawaiian Islands
1982 Jan 25	4.42	partial	(Antarctica)
Jun 21	12.03	partial	(Antarctica)
Jul 20	18.44	partial	(Arctica)
Dec 15	9.31	partial	(Arctica)
1983 Jun 11	4.42	total	SW Indian Ocean, Indonesia, New Guinea, Melanesia
Dec 4	12.30	annular	NW Atlantic, Gabon, Ethiopia, E Somali
1984 May 30	16.45	annular	C Pacific, N Mexico, Louisiana, Virginia, C Algeria
Nov 22	22.53	total	Moluccas, New Guinea, SE Pacific
1985 May 19	21.29	partial	(Arctica)
Nov 12	14.10	total	S Pacific, Antarctica
1986 Apr 9	6.20	partial	(Antarctica)
Oct 3	19.05	hybrid	Denmark Strait, N Atlantic
1987 Mar 29	12.49	hybrid	S Argentina, . . . Gabon, C Africa, Ethiopia, N Somalia
Sep 23	3.11	annular	USSR (Kazaeh), N China, E China Sea, Polynesia
1988 Mar 18	1.58	total	NE Indian Ocean, Indonesia, Borneo, S Philippines, Gulf of Alaska
Sep 11	4.43	annular	Somali coast, S Indian Ocean, SW Pacific
1989 Mar 7	18.07	partial	(Arctica)
Aug 31	5.31	partial	(Antarctica)
1990 Jan 26	19.30	annular	Antarctica, SE Atlantic
Jul 22	3.02	total	Finland, White Sea, N Siberia, NE Pacific
1991 Jan 15	23.53	annular	E Indian Ocean, S Western Australia, N Tasmania, New Zealand (Nelson & Wellington), E Pacific
Jul 11	19.06	total	C Pacific, Hawaii, Mexico, C America, W Colombia, C Brazil
1992 Jan 4	23.04	annular	Micronesia, Pacific, S California coast
Jun 30	12.10	total	S Uruguay, S Atlantic, SW Indian Ocean
Dec 24	0.30	partial	(Arctica)
1993 May 21	14.19	partial	(Arctica)
Nov 13	21.44	partial	(Antarctica)
1994 May 10	17.11	annular	C Pacific, N Mexico, USA (Texas, Okla., Mo., Ill., Ind., Ohio, Penn., N.Y., Mass.) . . . Morocco
Nov 3	13.40	total	SE Pacific, S Peru, S Brazil, S Atlantic, SW Indian Ocean
1995 Apr 29	17.32	annular	S Pacific, N Peru, Amazonia, C Atlantic
Oct 24	4.32	total	Iran, N India, Indo-China, N Borneo, Pacific (Marshall Is.)

Date	Time	Type	Central shadow belt passes through:
1996 Apr 17	22.37	partial	(Antarctica)
Oct 12	14.02	partial	(Arctica)
1997 Mar 9	1.23	total	N Tibet, Siberia, Arctica
Sep 2	0.03	partial	(Antarctica)
1998 Feb 26	17.28	total	C Pacific, N Polynesia, NW Colombia, Atlantic (Canary Is.)
Aug 22	2.06	annular	NE Indian Ocean, Sumatra, N Borneo, Melanesia, Tubnai Is.
1999 Feb 16	6.33	annular	SE Atlantic, S Indian Ocean, Australia (W.A., N.T., Qld., Coral Sea)
Aug 11	11.03	total	NW Atlantic, Britain (Land's End), N France, S Germany, Austria, S Hungary, Romania, Black Sea, N Turkey, Iran, India, Bay of Bengal
2000 Feb 5	12.49	partial	(Antarctica)
Jul 1	19.32	partial	(Antarctica)
Jul 31	2.13	partial	(Arctica)
Dec 25	17.34	partial	(Arctica)
2001 Jun 21	12.03	total	SW Atlantic, Angola, Zambia, N Mozambique, S Madagascar, SW Indian Ocean
Dec 14	20.51	annular	Pacific (Midway Is.), Costa Rica, Caribbean Sea
2002 Jun 10	23.44	annular	Celebes, Micronesia, N Pacific, W Mexico coast
Dec 4	7.31	total	SE Atlantic, Angola, Zimbabwe, S Mozambique, S Indian Ocean, Australia (south)
2003 May 31	4.08	annular/N	E Greenland, Iceland
Nov 23	22.49	total	S Indian Ocean, Antarctica
2004 Apr 19	13.33	partial	(Antarctica)
Oct 14	2.59	partial	(Arctica)
2005 Apr 8	20.35	hybrid	SW Pacific, N Colombia, E Venezuela
Oct 3	10.31	annular	N Atlantic, Spain, Algeria, S Somalia, NE Indian Ocean
2006 Mar 29	10.11	total	NE Brazil, . . . S Ghana, W Egypt, S & W Turkey, N Caspian Sea, N Mongolia
Sep 22	11.39	annular	Guyana, S Atlantic, S Indian Ocean
2007 Mar 19	2.31	partial	(Arctica)
Sep 11	12.31	partial	(Antarctica)
2008 Feb 7	3.54	annular	Antarctica, S Pacific
Aug 1	10.20	total	Canada (Franklin Dist.), N Greenland, W Siberia, China
2009 Jan 26	7.58	annular	SE Atlantic, S Indian Ocean, Indonesia, Borneo, Celebes Sea
Jul 22	2.35	total	N India, China, Micronesia, Polynesia
2010 Jan 15	7.06	annular	C African Rep., Kenya, S Somalia, . . . N Sri Lanka, Burma, China, Yellow Sea
Jul 11	19.33	total	S Polynesia, . . . S Chili

APPENDIX

Table 6.2 Lunar eclipses

Times given are the moment of ecliptical conjunction in UT

Date	Time	Type
1981 Jul 17	4.48	partial
1982 Jan 9	19.56	total
Jul 6	7.30	total
Dec 30	11.26	total
1983 Jun 25	8.25	partial
1984 –		
1985 May 4	19.57	total
Oct 28	17.43	total
1986 Apr 24	12.44	total
Oct 17	19.19	total
1987 Oct 7	3.59	partial
1988 Aug 27	11.06	partial
1989 Feb 20	15.37	total
Aug 17	3.04	total
1990 Feb 9	19.12	total
Aug 6	14.07	partial
1991 Dec 21	10.34	partial
1992 Jun 15	4.57	partial
Dec 9	23.43	total
1993 Jun 4	13.00	total
Nov 29	6.26	total
1994 May 25	3.28	partial
1995 Apr 15	12.17	partial
1995 Apr 15	12.17	partial

Date	Time	Type
1996 Apr 4	0.09	total
Sep 27	2.53	total
1997 Mar 24	4.41	partial
Sep 16	18.47	total
1998 –		
1999 Jul 28	11.36	partial
2000 Jan 21	4.44	total
Jul 16	13.55	total
2001 Jan 9	20.21	total
Jul 5	14.58	partial
2002 –		
2003 May 16	3.39	total
Nov 9	1.18	total
2004 May 4	20.30	total
Oct 28	3.04	total
2005 Oct 17	12.02	partial
2006 Sep 7	18.53	partial
2007 Mar 3	23.21	total
Aug 28	10.35	total
2008 Feb 21	3.27	total
Aug 16	21.07	partial
2009 Dec 31	19.25	partial
2010 Jun 26	11.36	partial
Dec 21	8.16	total
2010 Jun 26	11.36	partial
Dec 21	8.16	total

Table 6.3 *Conjunctions of Mercury*

Superior	Inferior	Superior	Inferior	Superior	Inferior
1980 Aug 26	1980 Nov 3	1980 Dec 31	1981 Feb 17	1981 Apr 27	1981 Jun 22
1981 Aug 10	1981 Oct 18	1981 Dec 10	1982 Feb 1	1982 Apr 11	1982 Jun 1
1982 Jul 25	1982 Oct 2	1982 Nov 19	1983 Jan 16	1983 Mar 26	1983 May 12
1983 Jul 9	1983 Sep 15	1983 Oct 30	1983 Dec 31	1984 Mar 8	1984 Apr 22
1984 Jun 23	1984 Aug 28	1984 Oct 10	1984 Dec 14	1985 Feb 19	1985 Apr 3
1985 Jun 7	1985 Aug 10	1985 Sep 22	1985 Nov 28	1986 Feb 1	1986 Mar 16
1986 May 23	1986 Jul 23	1986 Sep 5	1986 Nov 13 T	1987 Jan 12	1987 Feb 27
1987 May 7 *	1987 Jul 4	1987 Aug 20	1987 Oct 28	1987 Dec 23	1988 Feb 11
1988 Apr 20	1988 Jun 13	1988 Aug 3	1988 Oct 11	1988 Dec 1	1989 Jan 25
1989 Apr 4	1989 May 23	1989 Jul 18	1989 Sep 24	1989 Nov 10 *	1990 Jan 9
1990 Mar 19	1990 May 3	1990 Jul 2	1990 Sep 8	1990 Oct 22	1990 Dec 24
1991 Mar 2	1991 Apr 14	1991 Jun 17	1991 Aug 21	1991 Oct 3	1991 Dec 8
1992 Feb 12	1992 Mar 26	1992 May 31	1992 Aug 2	1992 Sep 15	1992 Nov 21
1993 Jan 23	1993 Mar 9	1993 May 16 *	1993 Jul 15	1993 Aug 29	1993 Nov 6 T
1994 Jan 3	1994 Feb 20	1994 Apr 30 *	1994 Jun 25	1994 Aug 13	1994 Oct 21
1994 Dec 14	1995 Feb 3	1995 Apr 14	1995 Jun 5	1995 Jul 28	1995 Oct 5
1995 Nov 23	1996 Jan 18	1996 Mar 28	1996 May 15	1996 Jul 11	1996 Sep 17
1996 Nov 1	1997 Jan 2	1997 Mar 11	1997 Apr 25	1997 Jun 25	1997 Aug 31
1997 Oct 13	1997 Dec 17	1998 Feb 22	1998 Apr 6	1998 Jun 10	1998 Aug 13
1998 Sep 25	1998 Dec 1	1999 Feb 4	1999 Mar 19	1999 May 25	1999 Jul 26
1999 Sep 8	1999 Nov 15 T	2000 Jan 16	2000 Mar 1	2000 May 9 *	2000 Jul 6
2000 Aug 22	2000 Oct 30	2000 Dec 25	2001 Feb 13	2001 Apr 23	2001 Jun 16
2001 Aug 5	2001 Oct 14	2001 Dec 4	2002 Jan 27	2002 Apr 7	2002 May 27
2002 Jul 21	2002 Sep 27	2002 Nov 14 *	2003 Jan 11	2003 Mar 21	2003 May 7 T
2003 Jul 5	2003 Sep 11	2003 Oct 25	2003 Dec 27	2004 Mar 4	2004 Apr 17
2004 Jun 18	2004 Aug 23	2004 Oct 5	2004 Dec 10	2005 Feb 14	2005 Mar 29
2005 Jun 3	2005 Aug 5	2005 Sep 18	2005 Nov 24	2006 Jan 26	2006 Mar 12
2006 May 18	2006 Jul 18	2006 Sep 1	2006 Nov 8 T	2007 Jan 7	2007 Feb 23
2007 May 3 *	2007 Jun 28	2007 Aug 16	2007 Oct 24	2007 Dec 17	2008 Feb 7
2008 Apr 16	2008 Jun 7	2008 Jul 30	2008 Oct 6	2008 Nov 25	2009 Jan 20
2009 Mar 31	2009 May 18	2009 Jul 14	2009 Sep 20	2009 Nov 5 *	2010 Jan 4
2010 Mar 14	2010 Apr 28	2010 Jun 28	2010 Sep 3	2010 Oct 17	2010 Dec 20

T = transit of Mercury across Sun's disk.
* = occultation of Mercury by Sun's disk.

APPENDIX

Table 6.4 Conjunctions of Venus

Lion	Bull	Fishes	Goat/Archer	Scales/Virgin
1979 Aug 25 *s*	1980 June 15 *i*	1981 April 7 *s*	1982 Jan 21 *i*	1982 Nov 4 *s*
1983 Aug 25 *i*	1984 June 15 *s* *	1985 April 3 *i*	1986 Jan 19 *s*	1986 Nov 5 *i*
1987 Aug 23 *s*	1988 June 12 *i*	1989 April 4 *s*	1990 Jan 18 *i*	1990 Nov 1 *s*
1991 Aug 22 *i*	1992 June 13 *s* *	1993 April 1 *i*	1994 Jan 17 *s*	1994 Nov 2 *i*
1995 Aug 20 *s*	1996 June 10 *i*	1997 April 2 *s*	1998 Jan 16 *i*	1998 Oct 30 *s*
1999 Aug 20 *i*	2000 June 11 *s* *	2001 March 30 *i*	2002 Jan 14 *s*	2002 Oct 31 *i*
2003 Aug 18 *s*	2004 June 8 *i* *T*	2005 March 31 *s*	2006 Jan 14 *i*	2006 Oct 27 *s*
2007 Aug 18 *i*	2008 June 9 *s*	2009 March 27 *i*	2010 Jan 11 *s*	2010 Oct 29 *i*

i = inferior conjunction   *s* = superior conjunction
T = transit of Venus across Sun's disk   * = occultation of Venus by Sun's disk.

Table 6.5 Visibility of Venus

years	months	time	max. brill.	visibility*
1979 1987 1995 2003	Jan – July	Morn	—	fair
1980 1988 1996 2004	Oct – Dec ⎫ Jan – May ⎭	Eve	beg. May	good
1981 1989 1997 2005	July – Dec ⎫ Jan ⎭	Morn	mid July	good
1982 1990 1998 2006	May – Dec ⎫ beg. Jan ⎭	Eve	mid Dec	bad
1983 1991 1999 2007	end Jan – Sep	Morn	end Feb	bad
	Dec ⎫ Jan – July ⎭	Eve	mid July	fair
1984 1992 2000 2008	Oct – Dec ⎫ Jan – April ⎭	Morn	Sep/Oct	good
1985 1993 2001 2009	July – Dec ⎫ Jan – April ⎭	Eve	end Feb	good
1986 1994 2002 2010	May – Dec	Morn	mid May	bad
	Feb – Nov	Eve	end Sep	bad
	Dec	Morn	mid Dec	fair

*For southern hemisphere visibility is opposite (good where bad is stated).

Table 6.6   Position of Mars

Position in constellations and dates of ecliptical conjunction and opposition, with ecliptical longitude. Best visibility is two months around opposition.

* = constellation where Mars makes a loop.

Year	Constellation	Conjunction	Opposition
1981	♑ – ♍	Apr 1 11°	–
1982	♍* – ♑	–	Mar 29 188°
1983	♑ – ♍	Jun 4 73°	–
1984	♍ ♎* – ♒	–	May 8 229°
1985	♒ – ♍ (♎)	Jul 19 116°	–
1986	♎ – ♐* – ♒ (♓)	–	Jul 6 284°
1987	♓ – ♎	Aug 26 152°	–
1988	(♎) ♏ – ♓ *	–	Sep 24 1°
1989	♓ – ♏	Sep 30 186°	–
1990	♏ – ♉*	–	Nov 26 63°
1991	♉ – ♏	Nov 10 227°	–
1992	(♏) ♐ – ♊*	–	–
1993	♊* – ♐	Dec 26 274°	Jan 6 105°
1994	♐ – ♌	–	–
1995	♌* ♋ – ♐	–	Feb 9 141°
1996	(♐) ♑ – ♍	Feb 29 340°	–
1997	♍* ♌ – ♑	–	Mar 15 175°
1998	♑ – ♍	May 13 52°	–
1999	♍ (♎) ♍* – ♑ (♒)	–	Apr 21 212°
2000	♒ – ♍	Jul 1 99°	–
2001	♍ – ♏* (♐) ♏ –	–	Jun 13 262°
2002	♒ – ♎	Aug 10 138°	–
2003	♎ – ♒* ♓	–	Aug 27 334°
2004	♓ – ♏	Sep 15 173°	–
2005	♏ – ♈*	–	Nov 7 45°
2006	♈ – ♏	Oct 23 210°	–
2007	♏ – ♊*	–	Dec 24 93°
2008	♉* – ♐	Dec 5 254°	–
2009	♐ – ♋ (♌)	–	–
2010	♌* ♋ – ♐	–	Jan 29 129°

Table 6.7 Position of Jupiter

Position in constellations and dates of conjunction and opposition, with ecliptical longitude.
Best visibility is two months around opposition.

Year	Constellation	Conjunction	Opposition
1981	♍	Oct 14 201°	Mar 26 185°
1982	♍ ♎ ♏	Nov 13 231°	Apr 26 215°
1983	♏	Dec 14 262°	May 27 246°
1984	(♏) ♐	–	Jun 29 278°
1985	♐ ♑	Jan 15 395°	Aug 4 312°
1986	♑ ♒(♓)	Feb 18 330°	Sep 10 348°
1987	♒ ♓ (♈)	Mar 27 6°	Oct 18 25°
1988	♓ ♈ ♉	May 3 43°	Nov 23 61°
1989	♉ ♊	Jun 9 79°	Dec 27 96°
1990	♊ ♋	Jul 15 112°	–
1991	♋ ♌	Aug 18 144°	Jan 29 129°
1992	♌ ♍	Sep 17 175°	Feb 29 160°
1993	♍	Oct 18 205°	Mar 30 190°
1994	♎ (♍) ♏	Nov 17 235°	Apr 30 220°
1995	♏ (♐)	Dec 19 267°	Jun 1 250°
1996	♐	–	Jul 4 283°
1997	(♐)♑	Jan 19 300°	Aug 9 317°
1998	(♑) ♒ ♓	Feb 23 335°	Sep 16 354°
1999	♓ ♈	Apr 1 11°	Oct 23 30°
2000	♓ ♈ ♉	May 8 48°	Nov 28 66°
2001	♉ ♊	Jun 14 83°	–
2002	♊ ♋	Jul 19 117°	Jan 1 101°
2003	♋ ♌	Aug 22 149°	Feb 2 133°
2004	♌ ♍	Sep 22 180°	Mar 4 164°
2005	♍ (♎)	Oct 22 209°	Apr 3 194°
2006	♎ ♏	Nov 22 240°	May 4 224°
2007	♏ ♐	Dec 23 271°	Jun 6 255°
2008	♐ (♑)	–	Jul 9 288°
2009	♑ ♒	Jan 24 305°	Aug 14 322°
2010	♒ ♓	Feb 28 340°	Sep 21 358°

Table 6.8   Position of Saturn

Position in constellations and dates of conjunction and opposition, with ecliptical longitude. Best visibility is two months around opposition.

gt N (gt S) = greatest ring opening north (south)   ∅ = ring invisible

Year	Constellation	Conjunction	Opposition	Ring
1981	♍	Oct 6 193°	Mar 27 186°	
1982	♍	Oct 18 205°	Apr 9 198°	
1983	♍ ♎	Oct 31 218°	Apr 21 210°	
1984	♎	Nov 11 229°	May 3 223°	
1985	♎ ♏	Nov 23 241°	May 15 234°	
1986	♏	Dec 4 252°	May 28 246°	
1987	♏	Dec 16 264°	Jun 9 258°	Feb 10, May 16, Nov 6, gt N
1988	♏ ♐	Dec 26 275°	Jun 20 269°	
1989	♐	–	Jul 2 280°	
1990	♐	Jan 7 287°	Jul 14 292°	
1991	♐ ♑	Jan 18 298°	Jul 27 303°	
1992	♑	Jan 29 309°	Aug 7 316°	
1993	♑ ♒	Feb 9 321°	Aug 19 327°	
1994	♒	Feb 21 333°	Sep 1 339°	
1995	♒ ♓	Mar 6 345°	Sep 14 351°	May 21, Aug 11, ∅
1996	♒ ♓	Mar 17 357°	Sep 26 4°	Feb 11, ∅
1997	♓	Mar 31 10°	Oct 9 16°	
1998	♓ ♈	Apr 13 23°	Oct 22 29°	
1999	♓ ♈	Apr 27 37°	Nov 5 43°	
2000	♈ ♉	May 11 51°	Nov 18 57°	
2001	♉	May 25 64°	Dec 3 71°	
2002	♉	Jun 9 79°	Dec 17 85°	Dec, gt S
2003	♉ ♊	Jun 25 93°	Dec 31 100°	
2004	♊	Jul 8 107°	–	
2005	♊ ♋	Jul 22 120°	Jan 14 114°	
2006	♋ ♌	Aug 7 135°	Jan 28 128°	
2007	♌	Aug 22 149°	Feb 10 142°	
2008	♌	Sep 4 162°	Feb 24 155°	
2009	♌ ♍	Sep 17 175°	Mar 9 169°	Sep 4, ∅
2010	♍	Oct 1 188°	Mar 22 182°	

APPENDIX

Table 6.9 Positions of Uranus, Neptune, Pluto

Uranus		Neptune		Pluto	
1981 – 1987	Scorpion	1969 – 1983	Scorpion	1968 – 1987	Virgin
1988 – 1995	Archer	1984 – 1997	Archer	1987 – 1993	Balance
1996 – 2002	Goat	1998 – 2010	Goat	1994 – 2015	Scorpion
2002 – 2007	Water-Bearer				
2008 – 2017	Fishes				

Conjunction of Uranus and Neptune: 1992 Feb 19, July 7, Nov 30 at 286°.

Bibliography

Ahnert, Paul, *Astronomisch-Chronologische Tafeln*, Ambrosius, Leipzig 1971.

Allen, Richard H., *Star Names*, 1899 (reprinted Dover, N.Y. 1963).

Astronomical Ephemeris, H.M.S.O. London, & Washington, annually.

Baravalle, Hermann von, *Astronomy, an Introduction*, Waldorf School/St George Books, New York 1974.

Bergh, G. van den, *Periodicity and Variation of Solar (and Lunar) Eclipses*, 2 vols., Willink, Haarlem 1955.

Bittleston, Adam, *The Seven Planets*, Floris, Edinburgh 1985.

Blattmann, Georg, *Comets, their Appearance and Significance*, Floris, Edinburgh 1985.

——, *The Sun, the Ancient Mysteries and a New Physics*, Floris, Edinburgh & Anthroposophic, New York 1985.

Davidson, Norman, *Astronomy and the Imagination*, Routledge & Kegan Paul, London 1985.

Falck-Ytter, Harald, *Aurora, the Northern Lights in Mythology, History and Science*, Floris, Edinburgh & Anthroposophic, New York, 1985.

Gingerich, O., *see* Stahlman, W. D., & O. Gingerich.

Grosjean, C., *see* Meeus, J., C. Grosjean, & W. Vanderleen.

Meeus, Jean, *Astronomical Tables of Sun, Moon and Planets*, Willmann, Bell, Richmond 1983.

Meeus, Jean, C. Grosjean, & W. Vanderleen, *Canon of Solar Eclipses*, Pergamon, Oxford 1966.

Meeus, Jean, *see also* Pilcher, Frederick, & Jean Meeus.

Muchery, Georges, *Tables des Positions Planétaires de 1937 à 2000* (6-day intervals) Chariot, Paris, n.d.

Oppolzer, Theodor von, *Canon der Finsternisse*, Wien 1887.

Pilcher, Frederick, & Jean Meeus, *Tables of Minor Planets*, Illinois College.

Stahlman, W. D., & O. Gingerich, *Solar and Planetary Longitudes for Years -2500 to 2000 by 10-day Intervals*, University of Wisconsin, 1973.

Sternkalender, Goetheanum, Dornach, annually.

Tuckerman, Bryant, *Planetary, Lunar and Solar Positions 601 BC to AD 1, and AD 2 to 1649*. American Philosophical Society, Philadelphia 1962.

Vanderleen, W., *see* Meeus, J., C. Grosjean, & W. Vanderleen.

Vreede, Elisabeth, *Astronomy and Anthroposophy*, Anthroposophic, New York 1987.

Webb, E. J., *The Names of the Stars*, London 1952.

Index

237

Floris Books

For news on all our **latest books,**
and to receive **exclusive discounts,**
join our mailing list at:

florisbooks.co.uk

Plus subscribers get a FREE book
with every online order!

We will never pass your details to anyone else.

Printed in the USA
CPSIA information can be obtained
at www.ICGtesting.com
JSHW05202530l024
72691JS00005B/29